Surfaces and Interfaces for Renewable Energy

Surfaces and Interfaces for Renewable Energy

Special Issue Editors

Francisco Manzano-Agugliaro
Aránzazu Fernández-García

MDPI • Basel • Beijing • Wuhan • Barcelona • Belgrade

Special Issue Editors
Francisco Manzano-Agugliaro
University of Almeria
Spain

Aránzazu Fernández-García
CIEMAT—Plataforma Solar de Almería
Spain

Editorial Office
MDPI
St. Alban-Anlage 66
4052 Basel, Switzerland

This is a reprint of articles from the Special Issue published online in the open access journal *Coatings* (ISSN 2079-6412) in 2019 (available at: https://www.mdpi.com/journal/coatings/special_issues/surf_inter_renew_energy).

For citation purposes, cite each article independently as indicated on the article page online and as indicated below:

LastName, A.A.; LastName, B.B.; LastName, C.C. Article Title. *Journal Name* **Year**, *Article Number*, Page Range.

ISBN 978-3-03928-128-2 (Pbk)
ISBN 978-3-03928-129-9 (PDF)

Cover image courtesy of Francisco Manzano-Agugliaro.

Contents

About the Special Issue Editors

Francisco Manzano-Agugliaro, full Professor at the Engineering Department in the University of Almeria (Spain), received his M.S. as Agricultural Engineer and Ph.D. in Geomatics at the University of Cordoba (Spain). He has published over 150 papers in JCR journals (https://orcid.org/0000-0002-0085-030X), H-index 26. His main interests are energy, sustainability, scientometrics, water, and engineering. He has supervised 25 PhD Thesis. His credentials include Vice Dean of Engineering Faculty (2001–2004); Director of Central Research Services (2016–2019); PhD Program Coordinator for Environmental Engineering (2000 to 2012); Greenhouse Technology, Industrial, and Environmental Engineering (from 2010); General Manager of Infrastructures (from 2019) at University of Almeria. He has received the following awards: top reviewer in Cross-Field—September 2019 (Web of Science), 2019 Outstanding Reviewer Award (Energies), 2019 Winner of the Sustainability Best Paper Awards.

Aránzazu Fernández-García, senior researcher at CIEMAT- Plataforma Solar de Almería (Spain), received her MS degree in Solar Energy in 2007 and the PhD degree in Environmental Engineering in 2013 at the University of Almería (Spain). She has developed her research activity in CIEMAT-PSA since 2002. She has published 35 papers in JCR journals (https://orcid.org/0000-0001-6044-4306), H index 16. She was Supervisor of 2 PhD students, 11 undergraduate students, 5 master students, and 3 pre-doctoral students. She is co-author of over 75 contributions to International Conferences. She has been involved in 19 EU and 3 Spanish founded I+D grants and has been scientific responsible of 25 cooperation agreements with industries. She is Site Manager of the laboratories for optical characterization and durability testing of solar reflectors at PSA (OPAC).

Editorial

Surfaces and Interfaces for Renewable Energy

Francisco Manzano-Agugliaro [1,*] and Aránzazu Fernández-García [2]

[1] Department of Engineering, University of Almeria, ceiA3, 04120 Almeria, Spain
[2] CIEMAT-Plataforma Solar de Almería, Ctra. Senés, 04200 Tabernas, Spain; afernandez@psa.es
* Correspondence: fmanzano@ual.es; Tel.: +34-950-015791; Fax: +34-950-015491

Received: 5 December 2019; Accepted: 6 December 2019; Published: 9 December 2019

Abstract: Energy is a growing need in today's world. Citizens and governments are increasingly aware of the sustainable use that must be made of natural resources and the great negative impact on the environment produced by conventional energies. Therefore, developments in energy systems based on renewable energies must be carried out in the very near future. To ensure their sustainability, they must be made of durable materials, and for this, the study of coatings is extremely important. This is also vital in systems based on solar energy, where the optical properties of the materials must be preserved as long as possible, and to this must be added the fact that they tend to be installed in very aggressive environments from the point of view of corrosion. Therefore, this special issue aims to contribute to the development of this challenge.

Keywords: solar energy; coatings; thin film; reflector; light trapping; concentrating solar thermal energy; reflectance corrosion

---

1. Introduction

The worldwide demand for electricity will grow to 50% in the next 20 years, mainly due to the increase in the world population, the generalization of electric vehicles as a form of transport and the boom in the battery market. However, this huge increase will be covered almost completely by renewable energy sources. The durability of renewable energy systems depends to a large extent on their surfaces. The improvement of coatings is one of the great challenges of the engineering and material science applied to these systems. This Special Issue will focus on the developments in this particular domain.

This Special Issue includes theoretical or practical issues of the following topics of interest, but are not limited to:

- Antireflective coatings;
- Antisoiling coatings;
- Corrosion resistance coatings;
- Increased optical properties (reflectance, absorptance, transmittance, and emittance);
- Surface treatment;
- Solar cells;
- Scanning electron microscopy;
- X-ray diffraction;
- Thin films;
- Polymers;
- Plastic coatings;
- Corrosion;
- Nanoparticles and nanotechnology;

- Titanium dioxide;
- Carbon nanotubes;
- Aluminum coatings;
- Paints;
- Composite materials;
- Environmental impact;
- Lifetime prediction;
- Accelerated aging methods; and
- Optical measurement techniques.

2. Statistics of the Special Issue

The authors' geographical distribution by country for the published papers is shown in Table 1, where it is possible to observe 27 authors from Spain and Germany.

Table 1. Geographic distribution by the country of author.

Country	Number of Authors
Spain	16
Germany	11
Total	27

3. Authors of this Special Issue

The authors of this special issue and their main affiliations are summarized in Table 2, where there are four authors on average per manuscript.

Table 2. Affiliations and bibliometric indicators for the authors.

Author	Main Affiliation	Country	Reference
Francisco Buendía-Martínez	CIEMAT-Plataforma Solar de Almería	Spain	[1,2]
Aránzazu Fernández-García	CIEMAT-Plataforma Solar de Almería	Spain	[1,2]
Florian Sutter	German Aerospace Center (DLR)	Germany	[1,2]
Loreto Valenzuela	CIEMAT-Plataforma Solar de Almería	Spain	[1]
Alejandro García-Segura	CIEMAT-Plataforma Solar de Almería	Spain	[1]
Johannes Wette	German Aerospace Center (DLR)	Germany	[2]
David Argüelles-Arízcun	CIEMAT-Plataforma Solar de Almería	Spain	[2]
Itziar Azpitarte	IK4-Tekniker	Spain	[2]
Gema Pérez	Rioglass Solar S.A.	Spain	[2]
Ceyhun Oskay	DECHEMA-Forschungsinstitut	Germany	[3]
Tobias M. Meißner	DECHEMA-Forschungsinstitut	Germany	[3]
Carmen Dobler	DECHEMA-Forschungsinstitut	Germany	[3]
Benjamin Grégoire	DECHEMA-Forschungsinstitut	Germany	[3]
Mathias C. Galetz	DECHEMA-Forschungsinstitut	Germany	[3]
Karmele Vidal	IK4-Tekniker	Spain	[4]
Estíbaliz Gómez	IK4-Tekniker	Spain	[4]
Amaia Martínez Goitandia	IK4-Tekniker	Spain	[4]
Adrián Angulo-Ibáñez	IK4-Tekniker	Spain	[4]
Estíbaliz Aranzabe	IK4-Tekniker	Spain	[4]
Sophie Gledhill	Fraunhofer Institute for Solar Energy Systems	Germany	[5]
Kevin Steyer	Fraunhofer Institute for Solar Energy Systems	Germany	[5]
Charlotte Weiss	Fraunhofer Institute for Solar Energy Systems	Germany	[5]
Christina Hildebrandt	Fraunhofer Institute for Solar Energy Systems	Germany	[5]
Nuria Novas	University of Almeria	Spain	[6]
Alfredo Alcayde	University of Almeria	Spain	[6]
Dalia El Khaled	University of Almeria	Spain	[6]
Francisco Manzano-Agugliaro	University of Almeria	Spain	[6]

4. Brief Overview of the Contributions to This Special Issue

If a brief representation of all the keywords of the articles of the Special Issue is made by means of a cloud of words, Figure 1 is obtained. Here, it is observed that the predominant keywords are Solar, Coating, Film, and Thin.

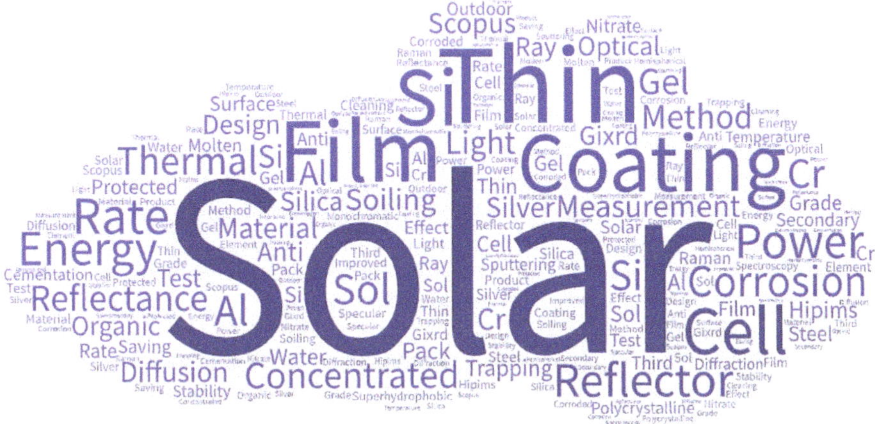

Figure 1. Cloudword of all the keywords.

Author Contributions: All authors contributed equally to this work.

Conflicts of Interest: The authors declare no conflict of interest.

References

1. Buendía-Martínez, F.; Fernández-García, A.; Sutter, F.; Valenzuela, L.; García-Segura, A. Advanced Analysis of Corroded Solar Reflectors. *Coatings* **2019**, *9*, 749. [CrossRef]
2. Wette, J.; Fernández-García, A.; Sutter, F.; Buendía-Martínez, F.; Argüelles-Arízcun, D.; Azpitarte, I.; Pérez, G. Water Saving in CSP Plants by a Novel Hydrophilic Anti-Soiling Coating for Solar Reflectors. *Coatings* **2019**, *9*, 739. [CrossRef]
3. Oskay, C.; Meißner, T.M.; Dobler, C.; Grégoire, B.; Galetz, M.C. Scale Formation and Degradation of Diffusion Coatings Deposited on 9% Cr Steel in Molten Solar Salt. *Coatings* **2019**, *9*, 687. [CrossRef]
4. Vidal, K.; Gómez, E.; Goitandia, A.M.; Angulo-Ibáñez, A.; Aranzabe, E. The Synthesis of a Superhydrophobic and Thermal Stable Silica Coating via Sol-Gel Process. *Coatings* **2019**, *9*, 627. [CrossRef]
5. Gledhill, S.; Steyer, K.; Weiss, C.; Hildebrandt, C. HiPIMS and DC Magnetron Sputter-Coated Silver Films for High-Temperature Durable Reflectors. *Coatings* **2019**, *9*, 593. [CrossRef]
6. Novas, N.; Alcayde, A.; El Khaled, D.; Manzano-Agugliaro, F. Coatings in Photovoltaic Solar Energy Worldwide Research. *Coatings* **2019**, *9*, 797. [CrossRef]

 *coatings* 

Review

Coatings in Photovoltaic Solar Energy Worldwide Research

Nuria Novas, Alfredo Alcayde, Dalia El Khaled and Francisco Manzano-Agugliaro *

Department of Engineering, ceiA3, University of Almeria, 04120 Almeria, Spain; nnovas@ual.es (N.N.);
aalcayde@ual.es (A.A.); dalia.elkhaled@gmail.com (D.E.K.)
* Correspondence: fmanzano@ual.es

Received: 23 October 2019; Accepted: 23 November 2019; Published: 27 November 2019

Abstract: This paper describes the characteristics of contributions that were made by researchers worldwide in the field of Solar Coating in the period 1957–2019. Scopus is used as a database and the results are processed while using bibliometric and analytical techniques. All of the documents registered in Scopus, a total of 6440 documents, have been analyzed and distributed according to thematic subcategories. Publications are analyzed from the type of publication, field of use, language, subcategory, type of newspaper, and the frequency of the keyword perspectives. English (96.8%) is the language that is most used for publications, followed by Chinese (2.6%), and the rest of the languages have a less than < 1% representation. Publications are studied by authors, affiliations, countries of origin of the authors, and H-index, which it stands out that the authors of China contribute with 3345 researchers, closely followed by the United States with 2634 and Germany with 1156. The Asian continent contributes the most, with 65% of the top 20 affiliations, and Taiwan having the most authors publishing in this subject, closely followed by Switzerland. It can be stated that research in this area is still evolving with a great international scientific contribution in improving the efficiency of solar cells.

Keywords: solar energy; coatings; scopus; material solar cell; thin film; polycrystalline; organic solar cell; thin film a-Si: H; optical design; light trapping

---

1. Introduction

Energy needs are a global growing problem in the era of technology. Citizens and governments are gradually becoming aware of the sustainable use of world resources [1]. Many are the developments in energy systems based on renewable energy, such as wind, photovoltaic, biomass, nuclear, etc., implemented on both a small and high scale. Renewable energy resources largely depend on the climate of the site; different renewable energies could be applied in different regions. Society demands clean and sustainable energy; this implies research in efficient clean energy. Among existing different renewable energies, solar energy is one of the most attractive for future energy sources [2–4] and photovoltaics is the most implemented one. Photovoltaic applications are very diverse, and they range from the incorporation into consumer products, such as watches, calculators, battery chargers, and a multitude of products from the leisure industry. They can also be applied in small-scale systems, like remote installations in structures, called solar gardens, or systems applied to the industrial and domestic facilities for small villages and water pumping stations. Not forgetting the large power production stations for supply of network connection. Energy policies play an important role in the development of renewable energy [5,6].

Currently, with the arrival of intelligent and sustainable buildings, solar modules that are installed in the building are installed in both roofs and part of the facade and windows, where, apart from energy efficiency, the aesthetics of the architecture are considered. Transparent and biphasic thin film solar modules contribute to their application in these structures [7].

These photovoltaic systems depend, to a large extent on the physical and chemical properties of their materials, the wavelength of the captured light, its intensity, and its angle of incidence, the characteristics of the surface or texture as well as the presence or absence of superficial coatings. In addition to these factors, temperature, pressure, ease of processing, durability, price, and costs throughout the life are important in material selection. Photovoltaic energy has been highly researched in the last 60 years, with the intention of reducing manufacturing costs and, at the same time, improving performance. The improvement of maintenance (protection against abrasion, corrosion, cleaning, etc.), increase in the life of the components, and incorporation of materials based on plastics and underlying substrates as coatings are among the cost reduction factors.

The starting silicon wafer is one of the main costs of silicon photovoltaic cells; the degree of purity largely defines the performance of the cell. This has led to the solar cells with nanostructure p-n radial junctions, where the quantity of Si and its quality is reduced. Improved light absorption in ultrafine solar silicon film is important in improving efficiency and reducing costs [8–10]. Thin-layer technologies also use less Si, reducing the production costs, although with limited efficiency, which increases the total system costs.

Research is being conducted for improving the capture of light in order to reduce the thickness of the layer, which entails reducing the material, and improving the efficiency, which has an impact on manufacturing costs. In this sense, solar cells have been improved by advances in diffractive optical elements (DOE) that are used in many areas of optics, such as spectroscopy and interferometry, among others. The shape of the grid slot can be used as an optimization parameter for specific tasks. In many cases, DOEs are manufactured on flat substrates for simplicity, but they offer many important additional advantages on curved surfaces [11]. With the development of computers and their application to different fields, such as homography, techniques such as interferometric recording have been developed. Digital homography (computer-generated holograms) has allowed for great flexibility in creating forms in substrates with high precision [12].

For this, nanostructures have been designed in different ways, depending on the type of solar cells. The compromise between optical and electrical performance currently limits solar cells. There are different proposals regarding whether nanostructures should be periodic or random. Non-fullerene acceptors (NFA) become an interesting family of organic photovoltaic materials and they have attracted considerable interest in their great potential in manufacturing large surface flexible solar panels through low-cost coating methods [13].

Research regarding the improvements in Solar Coating are in continuous evolution with the incorporation of new materials, structures, and the growing demand for energy; all these advances are mainly focused on improving the efficiency of photovoltaic panels. From this point of view, there are several scientific communities making continuous contributions from different fields. These contributions are doubled per decade, which entails a huge number of documents to deal with. The documents within the same field of research are distributed in scientific communities that are promoted through the interrelations between the authors and their publications. The collaboration of the authors in different communities makes the progress of science more productive, since there are not only research relationships between authors, but also between institutions that support the necessary tests with their laboratories and facilities. This exponentially increases the progress in science and technology. In this work, we study the different communities that have consolidated over time and the relationships between them.

2. Materials and Methods

This paper analyzes all of the scientific publications indexed on Scopus data base that deal with Solar Coating. There are search engines on the web based on Scientometric indicators, such as number and quality of contributions, according to the metric of the journal or the author. The results of these searches do not measure the relationships between the authors; this limits the establishment of collaborative communities. Technology, like science, advances through continuous collaborations between public or private research entities; therefore, it is important to develop metrics that incorporate the authors' relationships. There are different studies that carry out comparisons between Scopus and Web of Science, and they reach the conclusion that Scopus is the scientific database with the greatest contributions [14,15]. In addition, Scopus allows for the development of APIs (Application Programming Interface) that directly extract information from the database, allowing for an analysis of them [16]. Figure 1 shows the API developed, as it can be considered as the core of the methodology of this manuscript. Accordingly, a search for keywords related to Solar Coating has been carried out to find global relations between the generated communities, their authors, and research institutions. The search is performed for TITLE-ABS-KEY ("advanced glazing*" OR "Solar window*" OR "light trapping" OR "diffractive element*") obtaining many documents and their relationships. This requires a debugging process to avoid unnecessary information that prevents an overview, which reduces the number of documents and their relationships. Documents that have no relations within the generated communities are eliminated in the debugging process. The final data set was analyzed while using statistical tools that were based on diagrams and presentation of the data processed. The open source tool, like Gephi (https://gephi.org), was used, which incorporates statistical resources and data visualization, mainly the algorithm ForceAtlas2 [17]. In this way, the different clusters were automatically identified. After this, the information of each cluster was analyzed in Excel, while using the dynamic data tables and the word cloud has been realized with the software Word Art (https://wordart.com/create). Note that the size of the keyword must be proportional to its frequency and the number of times that keyword appears in the analyzed articles in a representation by cloud of words.

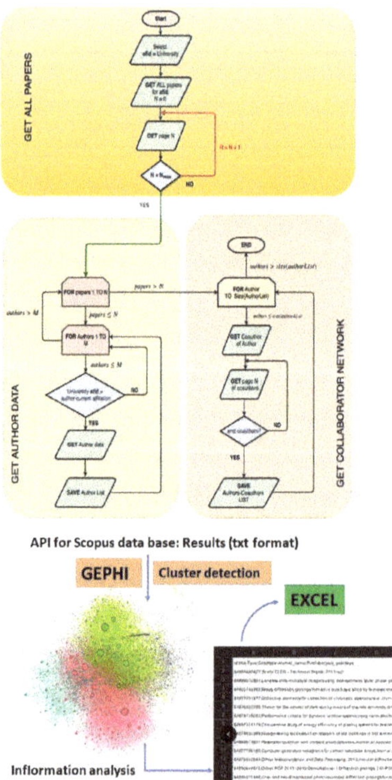

Figure 1. Flow diagram of the API that allowed for extracting the information of Scopus database.

3. Results

3.1. Communities Detection

A total of 6440 documents are obtained with a total of 21,301 relations between the authors after searching for the keywords. After the debugging process to avoid unnecessary information, documents are reduced by 39.1% and relations by 2.12%. Figure 2 shows the 3924 documents with 20,849 relationships that were obtained after the process of purification and statistical treatment. Figure 2 shows the distribution of the six detected communities that publish in Solar Coating topics with the Gephi program. As you can see, there is a main nucleus that is formed by five communities and another exterior formed by a single community. In Figure 2a, a node represents each publication and the size of the node is a function of their relationships, so that it shows the frequency with which the node appears in the shortest path between two randomly selected nodes in/between communities, showing the influence of the author within the community. In this way, not only the common metrics in search engines, such as Google Scholar, are considered, but also the collaborations between the authors. The size of a node varies according to its relationships to indicate the most influential nodes. The reason why an author who has a highly referenced and published document, but who works by himself, will only have a smaller node than a less referenced author with greater collaborations.

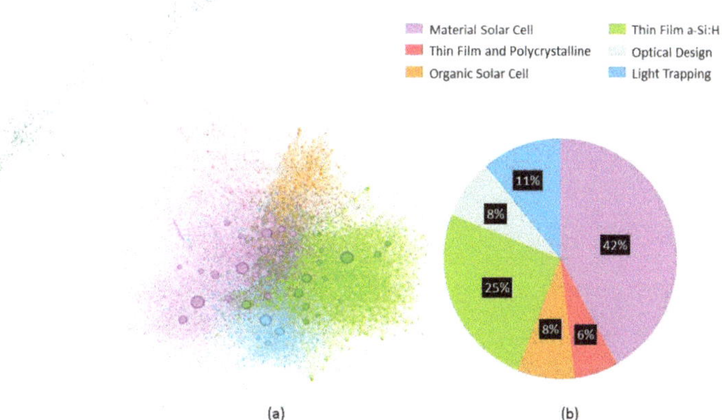

Figure 2. Representation of the communities investigating about "solar coating": (**a**) represent the interaction of the communities as a whole; and, (**b**) Representation for the distribution of the percentage of the communities.

Figure 2b presents the contribution in percentage of each community, since it is difficult to see the total size of each community due to the interrelation in Figure 2a. There are two communities that stand out for their size and they are the Material Solar Cell and Thin Fill Cells a-Si: H community. Community 0 (Material Solar Cell) is the largest with 42.2% of total publications. In this community, you can see the highest concentration of related nodes, where it publishes the advances on the materials used to improve the capture of light. The Thin Fill Cells a-Si: H community publishes 25.36% on the improvements in amorphous cells of hydrolyzed silicon. This community, besides being the second largest, is also the one that has a large concentration of authors that are related to other nodes, as it can be seen in Figure 2.

Figure 3 shows a cloud words of the global keywords obtained in the search. The most used keyword is "photovoltaic cells", with 147 times within the Material Solar Cell community, followed by "Thin film solar cells" with 96 times from the Thin film a-Si: H community. The third most used is "Silicon", also from the Solar Cell Material community with 85 times. The three words belong to the two communities with the largest number of publications, as shown in Figure 2b. Globally, the most repeated words are the most generic, such as "Film", "Thin", and "ZnO", as shown in Figure 3.

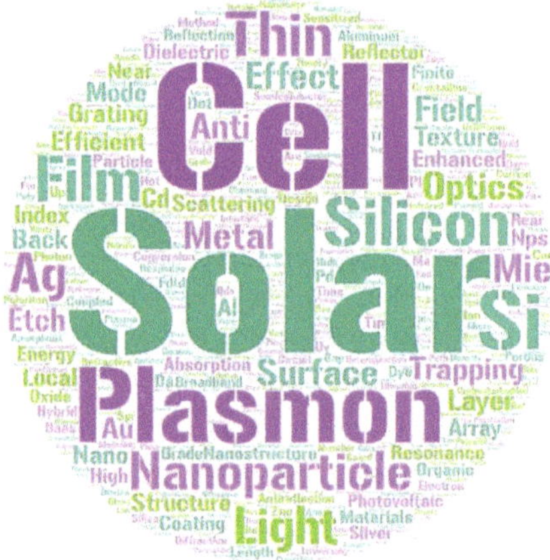

Figure 3. Cloud words of the keywords got in the global search.

3.2. Analysis of the Communities

Each community maintains a common theme, although being very interrelated with the rest of the communities. Next, the main nodes of each community and their most significant contributions are analyzed; this will allow for us to understand their research theme.

Community 0 (Material Solar Cell) investigates how to improve the capture of light by using different structures and materials used. Figure 4a shows the most representative keywords of the Solar Cell Material Community, which shows the number of times and their percentage of repetition within the community. The most representative word is "Photovoltaic cells", followed by "Silicon"; these words are very generic in the subject of photovoltaic solar energy and hints that this community is the widest when publishing more generic developments from which other more specific communities are nurtured. Hence, it has three important nodes that are very referenced, as shown in Figure 4b, and a multitude of publications that are very referenced not only by this community, but by the remaining ones, as it can be seen in Figure 2. Publications of the most referenced nodes in Scopus in order of size are:

- "Light trapping in silicon nanowire solar cells" [18] with a total of 1572 cites.
- "Fundamental limit of nanophotonic light trapping in solar cells" [19] with 586 cites.
- "Improving thin-film crystalline silicon solar cell efficiencies with photonic crystals" [20] with 532 cites.

Si cable assemblies are studied in this community. These are an interesting architecture for solar energy collection applications, and they can offer a mechanically flexible alternative to Si wafers for photovoltaic energy. Cables must absorb sunlight in a wide range of wavelengths and angles of incidence to achieve competitive conversion efficiencies, despite only occupying a modest fraction of the array volume. These matrices show a better near-infrared absorption, which allows for its absorption of sunlight to exceed the ray optics that traps the light absorption limit for an equivalent volume of textured plane, over a wide range of angles of incidence. The geometry of the cable network, together with other nanostructured geometries, offers opportunities to manipulate the relationship between the lighting area and the volume of absorption being useful in improving the efficiency or reducing the consumption of materials of many photovoltaic technologies. Kelzenberg et al. [21] showed that arrays presenting less than 5% of the cable area fraction can reach up to 96% of maximum absorption, and that they can absorb up to 85% of the substances integrated in the day, above the direct sunlight band. Garnett et al. [18] developed a structure of nanowires with large radial surface photovoltaic splicing p-n with efficiencies between 5% and 6%.

Material Solar Cell		
Keys	**Nº**	**%**
Photovoltaic cells	147	3.1
Silicon	85	1.8
Antireflection	29	0.6
Nanostructures	29	0.6
Photonic crystals	28	0.6
Silicon nanowires	26	0.6
Texturing	26	0.6
Photon management	25	0.5
Photonic cristal	23	0.5
Solar energy	23	0.5

(a)　　　　　　　　(b)

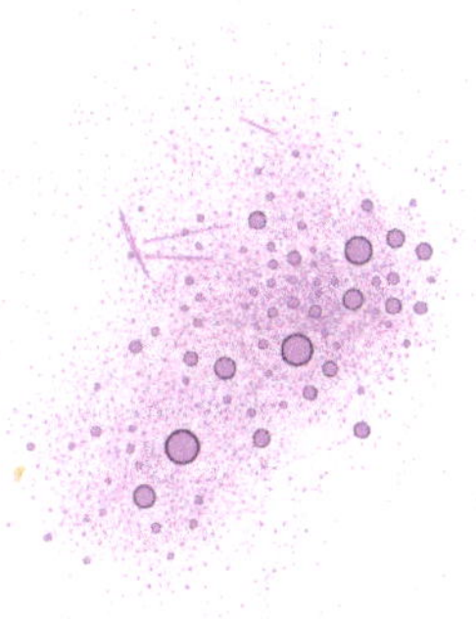

Figure 4. Representation of the Material Solar Cell community: (**a**) keywords; and, (**b**) isolated distribution of the publications.

Brongersma el al. [22] review the theory of nanophotonic light capture in periodic structures. Light collection schemes can be used to improve absorption in photovoltaic (PV) cells. They help to increase cell efficiency and reduce the production costs. In a homogeneous bulk cell with reflection mirror backing, (in a homogeneous bull cell with a back reflection mirror) the maximum enhancement factor attainable by the light trapping schemes is $4n^2/\sin^2(\theta)$, where n is the index of refraction of the material and θ is half of the apex angle of the absorption cone. Ultrafine cells with efficiencies that can exceed the traditional 4n2 limit are investigated. It involves the development of new computational tools that are capable of operating in the domain of wave optics, dealing with non-periodic structures and performing a joint electrical and optical optimization [23]. Yu et al. [19] studied the case of the capture of light in grid structures with periodicity at the wavelength scale. Light capture can improve cell efficiency, because thinner cells provide a better collection of photogenerated cells and potentially higher open circuit voltage. Yu et al. [19,24] developed "a statistical coupled-mode theory for nanophotonic light trapping" theory. Yu et al. [24], this theory is applied to the one-dimensional (1D) and two-dimensional (2D) grids that have close or even smaller thicknesses than the wavelength of the light and conclude that the 2D grids have a greater improvement factor. Yang et al. [25] used the coupled wave analysis method for textured sub length wavelength (STDS), which are important in obtaining high efficiency, due to their almost perfect anti-reflective properties.

Another author's study method was based on geometric optics and wave optics applied to thin-film crystalline silicon solar cell [20]. They manage to increase efficiency with the use of photonic glass, increasing 24.0% in an optimized 1D to 31.3% by adding an optimized 2D grid.

Wang et al. [26] present a double-sided grid design, in which the front and rear surfaces of the cell are separately optimized for antireflection and light capture, respectively. The authors propose a structure based on nano cones of different sizes for the upper layer (the period is 500 nm, the base radius is 250 nm, and the height is 710 nm) and lower layer (the period is 1000 nm, the base radius is

475 nm, and the height is 330 nm). Their experimental results approximate the limit of the theoretical absorption spectrum of Yablonovitch.

Community 1 (Thin Film and Polycrystalline) publishes the advances in the efficiency of thin cells and thin polycrystalline cells. Figure 5a shows the most representative keywords of the Thin Film and Polycrystalline community, which shows the number of times and their percentage of repetition within the community. The most representative word is "Silicon", followed by "Crystalline silicon" and "Epitaxy"; these words are generic of all communities. This community has a lot of keywords; it is the smallest and therefore its most repeated keywords are the most generic, the rest are more focused on the specific theme of the community. Figure 5b displays the distribution of published documents. Unlike the Material Solar Cell community, this is much more specific and, although it maintains connections with other communities, its articles do not have references from the Material Solar Cell and Thin film a-Si community: H.

Thin film & Polycrystalline		
Keys	**Nº**	**%**
Silicon	11	2.5
Crystalline silicon	7	1.6
Epitaxy	7	1.6
Absorption enhancement	5	1.1
Impurity photovoltaic effect	5	1.1
Photovoltaic	5	1.1
Polycrystalline silicon	5	1.1
Thin-film solar cells	5	1.1
Nanoimprint lithography	4	0.9
Porous silicon	4	0.9

(**a**)   (**b**)

Figure 5. Representation of Thin film and Polycrystalline community: (**a**) keywords; and, (**b**) isolated distribution of the publications.

In this community, a lot of research is being carried out to reduce the consumption of Si per watt peak. In addition to reducing the cost, a reduction in the thickness of the solar cell theoretically allows for an increase in the performance of the device. The long-term stability of thin film photovoltaic modules is increasing, while also reducing costs [27].

Thin film-based technologies show much lower surface production costs than bulk Si PV. Becker et al. [28] show the development of the i2 modules and the challenges that are faced by high-quality crystalline Si cells of thin film on glass, with the main ones being improvements in light trapping characteristics, low temperature junction processing, and cell metallization. Xue et al. [29] propose the Liquid Phase Crystallization Techniques (LPC) in the manufacture of high-quality crystalline silicon thin film solar cells in glass. Therefore, LPC is used for the development of double-sided silicon films, and different nanophotonic geometries of light capture are studied, concluding that this 10 mm thick double-sided silicon films can present maximum short-circuit current densities that are achievable in solar cells up to 38mA/cm^2 while assuming zero-parasite absorption.

Improvements in light entrapment in Si Polycristalline thin-layer solar cells (pc-Si) are based on the random scattering of light in the absorbent layer by glass substrate texture, or silicon film etching texture and plasmonic nanoparticles [29]. The thin-layer solar cells pc-Si on glass offer the possibility of achieving efficiencies of a single union of 15%. This is achieved by developing structures that improve light entrapment, being mainly based on silicon nanostructures, such as porous silicon, nanowires of silicon, and nano-silicon holes [29]. [30,31] proposes using a "seed layer" to obtain a high quality material, the use of ZnO and aluminum cladding as a method of improving light collection, and the use of high quality materials for the evaporation of the electron beam for the deposition of the absorbers, which offers a high potential for cost reduction, to obtain efficiency improvements and a reduction in costs. Another proposal is to use nanowire matrices to improve light entrapment and the design of the cell structure to minimize parasitic absorption, together with suppression of surface recombination, while using a multi-HIT configuration (hetero junction with intrinsic thin layer) core-based solar-based nanowire cells that were prepared in the thin film of low-cost pc-Si, developing an 8 μm pc-Si cell [32].

Another method that is based on surface plasmonic resonance (SPR) and a periodic hybrid matrix composed of a graphene ring at the top of the absorbent layer separated by an insulating layer to achieve an improvement of multiband absorption, increases the basis for simultaneous photodetection at multiple wavelengths with high efficiency and tunable spectral selectivity [33].

In [34], they propose a complete method for studying long-term light entrapment, the use of quantum efficiency data, and expressions of the calculation of Z0 and RBACK (reflectivity of the rear reflector defined in [10] for any solar cell), where Z0 is the optical path of short band length factor Z0 of Rand and Basore [35], and it is a multiple of the thickness of the cell necessary to generate equal to that found in the device. Although there are not very relevant nodes as compared to the others, it should be noted that the publications of the most cited nodes in Scopus for this scientific community in order of size are:

- "Polycrystalline silicon thin-film solar cells: Status and perspectives" [36] with a total of 117 cites.
- "Crystalline thin-foil silicon solar cells: Where crystalline quality meets thin-film processing" [37] with 64 cites.
- "Double-side textured liquid phase crystallized silicon thin-film solar cells on imprinted glasswith" [38] 37 times cited.

Community 2 (Organic solar cells) investigates an alternative to silicon-based photovoltaic cells, organic solar cells (OSC), or also called organic photovoltaic cells (OPV). Figure 6a shows the most representative keywords of the Organic solar cell's community, which shows the number of times and their percentage of repetition within the community. One of the most representative words is "Organic solar cells", after which the community is named. The following words are specific to the topic treated in this community, such as: "Organic photovoltaics", "Light harvesting", and "Polymer solar cell". Despite the small representativeness, 8.3% of the publications, (Figure 2), this community has a greater concentration of publications with references, as it can be seen in the size of the circles in Figure 6b, which is unlike the community Thin Film and Polycrystalline. This community has ties with the rest of the communities.

OSC cells have interesting advantages due to their characteristics, such as their lightweight, flexibility, and possibility of producing them profitably for large surfaces. These features have made of these cells very valid for applications in electronic textiles, synthetic leather, and robot, etc. The main disadvantage is their low energy conversion efficiency, which is mainly because the light absorption properties in an organic active layer have short optical absorption lengths ($L_A \sim 100$ nm) and exciton diffusion length ($L_D \sim 10$ nm). This implies that a reduction in the thickness of the active layer affects deterioration in performance, but an increase in thickness implies an increase in the series resistance and reduction in the collection of carriers. Therefore, a compromise between both of the situations is sought, efficient light collection and efficient load collection. The optical optimization that is used in other thin-layer technologies can be useful in achieving' maximum concentration in the absorbent layer, some of the proposals for improvement in light capture are based on modifying the structure, mainly plasmonic nanostructures, where photonic crystals are used, metal gratings, buried nanoelectrodes, etc.; however, in essence, they increase the organic surface layer [39,40]. Ko et al. [41] study the different nanostructure-based uptake systems for OSC cells while using both Plasmon surfaces and anti-reflective coatings, and photonic crystal (PC) nanostructure. Although theoretical calculations suggest that the efficiency increases the accumulation of absorption of light in nanostructured devices, the results show that they are still inferior to the highest reported conventional organic photovoltaic cells, which implies that further research in this field must be carried out to obtain thinner layers of photoactive material that improve the performance. The use of metallic nanomaterials can improve the capture of light in OSC and, although most nanoparticles (NPs) limit the improvement of the efficiency of power conversion to a narrow spectral range, broadband capture is desirable. The proposal of Li et al. [42] is the combination of Ag nanomaterials in different ways (Localized plasmonic resonances (LPRs), Ag nano prisms, and NPs mixed with Ag) for better broadband absorption and increased short-circuit photocurrent density. Out of the three experiments, the one with NPs mixed with Ag is the one with the highest efficiency with power conversion efficiency of 4.3%. They conclude that the cooperative plasmonic effects in metallic nanomaterials with different types of materials, shapes, size, and even the polarization incorporated in the active layers or between the layers or both should be further studied.

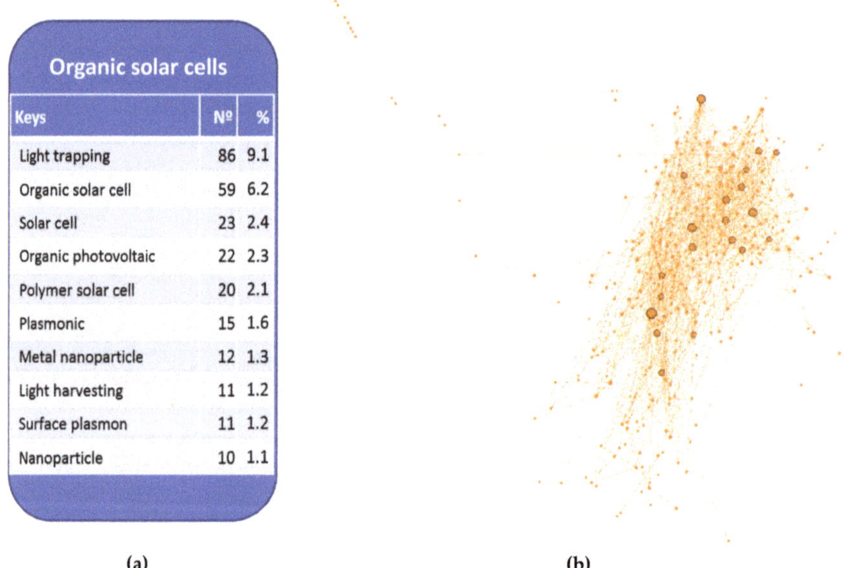

(a)	**(b)**

Figure 6. Representation of Organic solar cells community: (**a**) keywords; and, (**b**) isolated distribution of the publications.

The review that was carried out by Gan et al. [43] proposes incorporating plasmonic nanostructures in the front and rear metal electrodes of an OSC, which is expected to reach broadband, polarization, and absorption independent of the angle, and this implies the possibility of exceeding 10% power conversion efficiency. Tvingstedt et al. [44] analyzes the use of micro lens to increase the capture of light and, thereby, improves the absorption rate of the solar cell. Xiao et al. [45] propose a hybrid system of micro lens for OSC (a matrix of hybrid micro lens, a mirror with a matrix of holes, and an OSC with a reflective cathode) to improve broadband absorption. Each isana chromatic hybrid refractive-diffractive singlet micro lens made of a single optical material, and these hybrids micro lens are separated from the cells to avoid direct contact with an organic layer that can cause electrical defects. Another proposal is the use of a V-shaped light capture configuration; the purpose is to increase the photocurrent for all angles of incidence. Rim et al. [46] tested in a 170 nm polymer thin film OSC, obtaining a 52% improvement in efficiency and conclude that this V structure in thin film OSC is effective for active layer thicknesses of the order of wavelength of light or less.

Müller-Meskamp et al. [47] study direct patterning interfering laser (DLIP) has been used to manufacture periodic surface patterns (substrates with a 4.7 µm and hexagonal line of 0.7 µm) large surface area on flexible polyethylene terephthalate (PET) substrates. The results are encouraging, achieving the best results for the hexagonal corrugated structures with greater short-circuit current (Jsc) and greater energy conversion efficiency (PCE). Other important studies should be cited for Perovskite Solar Cells related to High-Performance Solution-Processed Double-Walled Carbon Nanotube Transparent Electrode [48], and the Highly reproducible perovskite solar cells with an average efficiency of 18.3% and best efficiency of 19.7% being fabricated via Lewis base adduct of lead (II) iodide [49]. Publications of the most cited in Scopus in this cluster are:

- "Plasmonic-enhanced organic photovoltaics: Breaking the 10% efficiency barrier" [50]. Cited 391 times.
- "An effective light trapping configuration for thin-film solar cells" [46]. Cited 170 times.
- "Light manipulation in organic photovoltaics" [51]. Cited 23 times.

Community 3 (Thin film a-Si: H) presents advances on thin-layer solar cells of hydrogenated Amorphous (a-Si:H) or hydrogenated microcrystalline (µc-Si:H). Figure 7a shows the most representative keywords of the Thin film a-Si: H community, which shows the number of times and their percentage of repetition within the community. The most representative word is "Thin film solar cells", which makes part of the community name, the next word is "Silicon", which is generic, and the rest are already more specific to the topic treated in this community, such as " Amorphous silicon "and" Microcrystalline silicon ". This community, although its studies are focused on a-Si cells: H is the second community in relation to total publications (Figure 2). It has a central core with a large number of references and three somewhat lower, but considerable references (Figure 7b). In addition, it maintains a great interaction with the rest of the communities that supply it with references. The publications of the most referenced nodes in Scopus in the order of size are the following three:

- "TCO and light trapping in silicon thin film solar cells" [52] with 869 cites.
- "Light trapping in solar cells: Can periodic beat random?" [53] was cited 369 but its node is large because it relates to major nodes.
- "Light trapping in ultrathin plasmonic solar cells" [54] had 512 cites, but with lower relations to major nodes.

The thin-layered Si (H-Si:H) or hydrogenated microcrystalline (µc-Si:H) thin-layer solar cells use an intrinsic layer (layer i) without doping between two highly doped layers (p and n). The optical and electrical properties of the i-layers are linked to the microstructure and, therefore, to the deposition rate of the layer i, which in turn affects the production yield [55]. The importance of contact and reflection in these solar cells require techniques to improve light uptake [52]. An integral part of these devices is the transparent conductor oxide (TCO) layers used as a front electrode and as a part of the rear side reflector [56]. When applied on the front side, the TCO must have high transparency in the spectral region, where the solar cell operates with high electrical conductivity. In p-i-n configuration, where the Si layers are deposited on a transparent substrate covered by TCO, with rough surfaces are applied in combination with the highly reflective rear contacts. TCO must have a strong dispersion of the incoming light in the silicon absorbent layer and favorable physicochemical properties for silicon growth. The application of zinc oxide films that were doped with aluminum (ZnO:Al) as a rear reflector result in a highly promising TCO material [57]. These films provide efficient coupling of the incident sunlight by refraction and light scattering at the interface TCO/Si to increase the length of the light path [52]. Another proposal for improving light entrapment is to use a return reflector; the use of Ag plasmonic nanoparticles can provide performance that is comparable to random textures in amorphous silicon solar cells n-i-p [58].

Thin film a-Si:H		
Keys	Nº	%
Thin film solar cells	97	3.0
Silicon	40	1.2
Amorphous silicon	39	1.2
Microcrystalline silicon	32	1.0
Back reflector	31	1.0
ZnO	28	0.9
Light scattering	26	0.8
Photovoltaics	26	0.8
Zinc oxide	25	0.8
Plasmonics	20	0.6

(**a**)   (**b**)

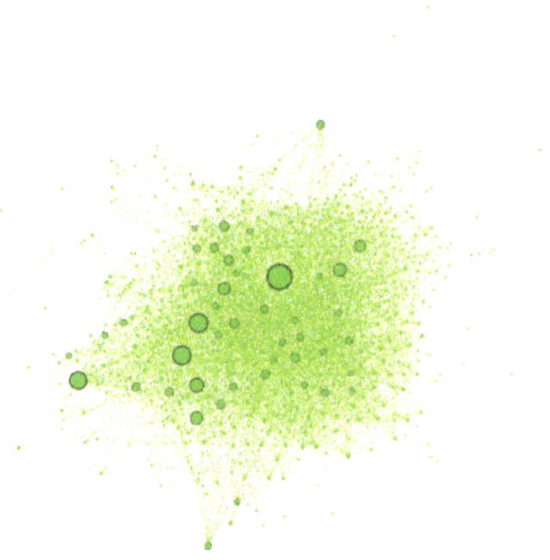

Figure 7. Representation of Thin film Si:H community: (**a**) keywords; and, (**b**) isolated distribution of the publications.

Other researchers propose different nanostructures for improving light entrapment, which are very important in thin-film amorphous silicon solar cells. Battaglia et al. [53] compare a random pyramidal nanostructure of transparent zinc oxide electrodes and a periodic one of periodic glass nanocavity matrixes manufactured by nanosphere lithography. The results show that both options have approximately a short-circuit current density of 17.1 mA/cm^2 and a high initial efficiency of 10.9%. Waveguide Theory provides a mechanism to select the period and symmetry of the grid to obtain efficiency improvements. The relationship between the photocurrent and the spatial correlations of random surfaces have been proposed by [59], developing pseudo-random matrices of nanostructures that are based on their power spectral density, and their correlation between the frequencies and the photocurrent.

Another option is nanodome solar cells, which have periodic nanoscale modulation for all types of solar cells from the lower substrate, through the active absorber to the upper transparent contact. These devices combine many nanophotonic effects to efficiently reduce the reflection and improve absorption over a wide spectral range. Nanodome solar cells with only one layer of 280 nm thick hydrogenated amorphous silicon (a-Si: H) can absorb 94% of the light with wavelengths of 400–800 nm, which is significantly greater than 65% absorption of flat film devices. In [60], they propose a nanodome solar cell of union p-i-n a-Si: H. The cells are composed of 100 nm thick Ag as a rear reflector, 80 nm Thick transparent conduction oxide (TCO) as a bottom and upper electrode, and a thin active layer of a-Si: 280 nm H (top to bottom): pin, 10-250-20 nm). Ferry et al. [54] proposed a strategy that consists on the use of non-randomized nanostructured reflectors optimized for ultra-thin solar cells of hydrogenated amorphous Si (a-Si:H). This alternative increases the short-circuit current densities, which improves the results as compared to cells that have posterior contacts with a flat or random texture.

Community 4 (Optical design) works on improvements in the efficiency of solar cells from the point of view of optical design by creating nanostructures to improve the capture of direct and diffused light. Figure 8a shows the most representative keywords of the optical design community, which shows the number of times and their percentage of repetition within the community. All of the words are very representative of the community, such as "Diffractive Optics" or "Optical Design", after which the community is named and it represents 11.5% each with respect to the community; the next words are "Diffractive Optics elements" and "Diffractive elements." This is the second smallest community, although it maintains links with the rest of the communities (Figure 2). This community incorporates publications from other disciplines, such as optics, which were not initially developed for solar energy, but whose impact on the optical behavior of a surface has been referenced by the other communities. In Figure 2, it appears as an emerging community. Unlike the other communities, it does not have a main nucleus, since this community does not have a great concentration of relationships in its articles (Figure 8b).

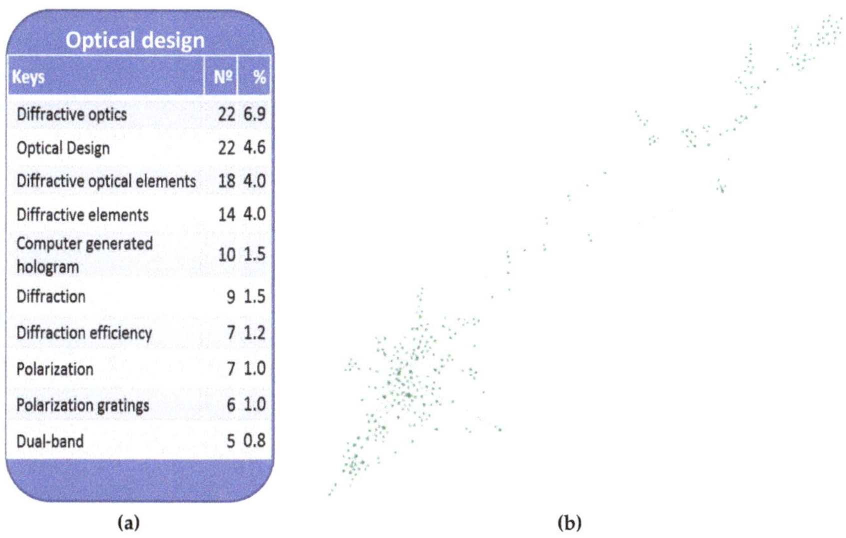

Figure 8. Representation of Optical design community: (**a**) keywords; and, (**b**) isolated distribution of the publications.

The manufacture of solar cells requires a prior study of their optical behavior with the consideration of better light capture. Therefore, their behavior is studied as a diffractive element, and lithographic structures respond differently, depending on their structure, composition, and size. Herkommer et al. [61] show simulation techniques to evaluate the distraction efficiency prior to the manufacture of the solar cell. It is very important prior to manufacturing to simulate the behavior in real conditions, since different manufacturing systems can be used, depending on the wavelength range of the incident and its size characteristics. Depending on the wavelength, more than one material can be used to implement the grid structures. The dependence on the size of the diffraction efficiency characteristic can be considered as the manufacturing limitations. There is commercial software that calculates the diffraction efficiency for a given grid while considering the period, grid shape, wavelength, and angle of incidence.

The diffractive optical element (DOE) can be implemented on flat and curved surfaces [12]. Grilles can be designed and implemented more complexly according to the need for the range of light collection and its efficiency with the use of direct laser lithography or through computer-generated holograms. On curved surfaces, they can be manufactured while using single-point diamond turning [62] to reduce the material used.

Digital holography (computer-generated hologram) has improved its quality in recent years, leading to unthinkable implementations for its accuracy not many years ago. Patterns can be engraved on photosensitive materials at the appropriate scale, embossed on high precision materials in various directions with or without periodicity, with the arrival of the laser and its computer control. Some lithographic techniques combine the coating with a light sensitive film (photo-resistance). Digital holography and Computer-Generated Holograms reduce the choice of material and pattern generation scheme. The degree of freedom in the choice of parameters limits the choice of coding technique and its optimization. The coding allows for adapting the data to the existing hardware requirements [12].

Solar cells have benefited from the study of the behavior of diffractive elements applied to other fields of research. The most common diffractive elements are a diffractive lens, a matrix generator, and a correlation filter [63]. The design is based on the optimum performance of the optical system and its manufacturing restrictions. These diffractive elements have a disadvantage in that they produce chromatic aberrations. Some authors [64] propose the attachment to the lens of a corrective substrate of the diffraction for application in the headlights of a car and study these elements for the diffractive telescope system.

Hybrids are one of the most complex diffractive elements, where they are both reflective and dissipative throughout the visible band (400–700 nm). Designing achromatic refractive-diffraction hybrid lenses is complex and it requires prior study for its manufacture to optimize its efficiency in the capture range [65]. Publications regarding the most cited in Scopus in this cluster are:

- "Digital holography as part of diffractive optics" [66] cited 111 times.
- "Understanding diffractive optic design in the scalar domain" [67] cited 110 times.
- "I digital holography computer-generated holograms" [68] cited 23 times.

Community 5 (Light trapping) investigates the improvement of the capture of light in solar cells, while using nanoparticles for the purpose of improving efficiency. Although many authors investigate and test systems for improving the light entrapment by nanostructures, in their turn they produce an increase in surface area and, therefore, minority recombination that reduces the efficiency. Other researchers use the dispersion of metal nanoparticles by varying their shape, size, particle material, and ambient dielectric energy to determine the improvement of light capture with particle plasmons to avoid this adverse situation. Figure 9a shows the most representative keywords of the Light trapping community, which shows the number of times and their percentage of repetition within the community. All of the words are very representative of the community as "Plasmonic" or "Surface plasmons", and they represent 4% and 3.1%, each with respect to the community, other words, like "Nanoparticles", "Silver nanoparticles", and "Metal nanoparticles", are very specific to the community, although within the 10 most repeated words, there are other generics, such as "Photovoltaics" and "Silicon" with 2.5% and 1.6% each. This community is ranked third in size (Figure 2) and it has a main nucleus and a somewhat smaller one as seen in Figure 9b. It is a community that is closely related to the rest, since the main topic discussed affects the studies of other communities, such as the entrapment of light, which, regardless of the type of cell, is important to improve and increase the efficiency. The two publications of the most referenced nodes in Scopus in the order of size are the following:

- "Surface plasmon enhanced silicon solar cells" with 1467 references and many relations with other communities [69].
- "Design principles for particle plasmon enhanced solar cells" [70] with 673 cites.
- "Tunable light trapping for solar cells using localized surface plasmons" [71] with 460 cites.

Catchpole et al. [70] show that the shape of the particles influences the path length, with the spherical shapes being worse than the cylindrical and hemispherical shapes. In addition, they conduct experiments with silver and gold particles, where the results show that those of silver provide a longer path length than those of gold. In Ouyang et al. [9], the effects of silver nanoparticles on polycrystalline silicon thin film solar cells on glass are studied, obtaining an improvement in the short-circuit current of a 1/3 increase when compared to conventional ones. The geometry of the matrix used also defines the entrapment of light (there are studies on random, quasi-periodic, and periodic matrices). Mokkapati et al. [72] study the behavior of silver nanoparticles in a periodic matrix and conclude that there is a very restrictive relationship between the optimal particle size and grid parameters of the periodic matrix; the case of silver particles of 200 nm, a 400 nm step is ideal for Si solar cells. Other authors analyzed the behavior of silver nanoparticles on the rear structure to reduce the entrapment losses that were produced with the long wavelength that escape (rear reflector dispersion), but with minimal electronic losses due to recombination effects. The photocurrent with the silver nanoparticles of a PERT (Passivated Emitter and Rear Totally Diffused) cell increases by 16% when compared to an aluminum rear structure [73].

Light trapping

Keys	Nº	%
Light trapping	104	7.9
Solar cells	65	5.0
Plasmonics	53	4.0
Surface plasmons	41	3.1
Photovoltaics	33	2.5
Nanoparticles	28	2.1
Silver nanoparticles	23	1.8
Silicon	21	1.6
Thin film solar cells	21	1.6
Metal nanoparticles	11	0.8

(**a**) (**b**)

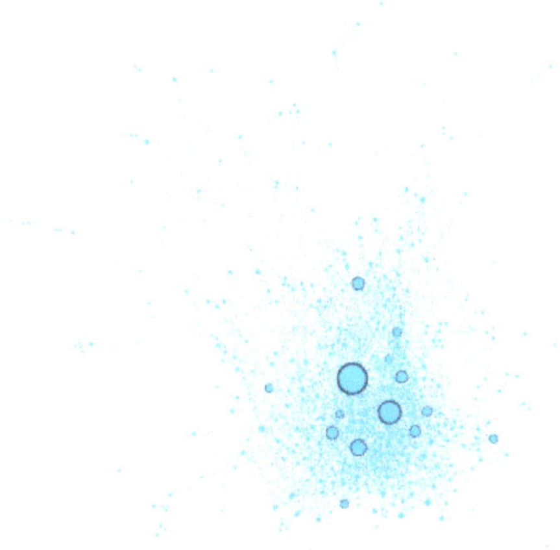

Figure 9. Representation of Light trapping community: (**a**) keywords, and (**b**) isolated distribution of the publications.

3.3. Analysis as Per Authors, Affiliations, and Countries of Investigation on Solar Coating

In total, there are 14,849 authors who research in 26 by subject area. The ten countries with the highest concentration of researchers contributing their scientific publications to the progress in Solar Coating have been analyzed. China makes the main contribution, with 3345 researchers (22.5%), being closely followed by the United States with 2634 (17.7%) and Germany with 1156 (7.8%). The rest is around 349 from Italy to 687 in South Korea, with percentages ranging from 2.4 to 4.6%, respectively. China and the United States are among the countries with the greatest contributions in scientific developments in all matters related to energy, and also in energy saving [16]. Figure 10 shows the distribution and percentages of participation of the authors according to the country of origin shown by colors. There are 28.8% of other countries of low percentage contribution, as can be seen in Figure 10, where Spain, Australia, and others appears.

The results show that China and the United States both actively collaborate in the progress by establishing collaborative ties with other countries in order to move forward and this can be seen from the intertwining of the communities in Figure 2. The main collaborations are established in Asia between Chinese, Koreans, Taiwanese, Japanese, and Indians, making a total of 38%. The European contributions between Germany, France, United Kingdom, and Italy make up for 16%. America is only represented by the United States.

Going deeper into the origin of the authors, the 20 most localized affiliations in the search have been analyzed, since it is important to know in which research and development centers the main contributions to Solar Coating are made. Table 1 shows the data; 12 of the 20 belong to the Asian continent with a contribution of 65% of the Top 20 publications. It is interesting how Switzerland, which was not among the 10 countries with the highest productivity in Solar Coating, is the second research institution in terms of the number of contributions with 6.6% of the top 20. In Europe, Germany stands out with the centers of the Fraunhofer Institute for Solar Energy Systems ISE and Helmholtz-Zentrum Berlin für Materialien und Energie (HZB) with a total of 11.1%, which together with Switzerland contribute in 17.7%. America contributes in 12.6% with the USA and only three research institutions, being the one that Stanford University publishes the most.

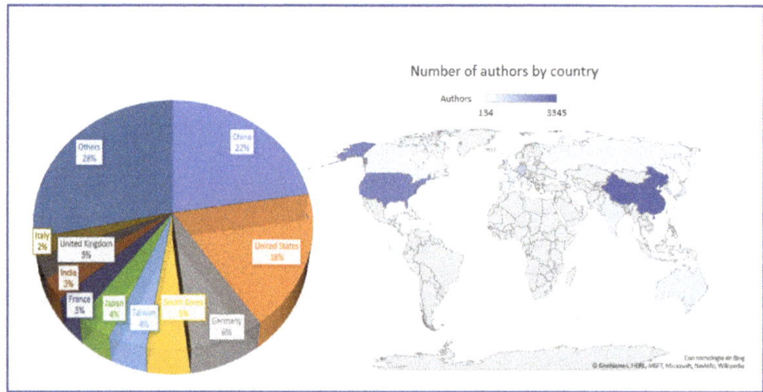

Figure 10. Author distribution per countries.

Table 1. Top 20 affiliations in Solar Coating.

Affiliation	Country	Publications
National Taiwan University	Taiwan	123
Swiss Federal Institute of Technology, Lausanne	Switzerland	115
Fraunhofer Institute for Solar Energy Systems ISE	Germany	108
Shanghai Jiao Tong University	China	105
Nankai University	China	103
Soochow University	China	103
Chinese Academy of Sciences	China	99
University of New South Wales (UNSW)	Australia	91
Forschungs zentrum jülich (FZJ)	Germany	84
Helmholtz-Zentrum Berlin für Materialien und Energie (HZB)	Germany	83
Stanford University	USA	82
Nanjing University	China	79
National Chiao Tung University	China	77
Jilin University	China	76
Sungkyunkwan University	South Korea	74
Nanyang Technological University	Singapore	69
National Renewable Energy Laboratory	USA	69
Massachusetts Institute of Technology	USA	67
Australian National University	Australia	64
Sun Yat-Sen University	China	62

Table 2 shows a list of the 10 authors with the highest H-index in Scopus in relation to the search performed. The order of the authors does not correspond to the order by institution of Table 1. The author with the highest H-index of 230, 1471 published documents, and 258,107 citations belong to Switzerland and there is a second in position 9 both same affiliations appearing in Table 1 (Swiss Federal Institute of Technology, Lausanne). In the top 10 by H-index, there are mainly US authors from Harvard University, Georgia Institute of Technology, Stanford Linear Accelerator Center, and University of California (Berkeley); none correspond to the three affiliations in Table 1. There are two Chinese authors in positions 8 and 10, with H-indexes of 138 and 130, respectively, which have a number of significant citations of 77,076 and 72789, in both cases their center does not correspond to those of Table 1. There is a German author with H-index of 186 and 136,496 citations, but their affiliation does not correspond to the affiliations of Germany in Table 1. It is significant that only the affiliation of the authors of Switzerland corresponds within those seen in Table 1. This might be due to the fact that, although the group of contributions has great references, it does not establish collaborations between the communities, and that means that the software debugging process has been eliminated. Figure 11 shows the location of the author of the highest H-index in Table 2. As it can be seen in the figure, it corresponds to the border that has not been represented when analyzing the communities, since it remains as a node with no connection to the rest. This can be interpreted that the system correctly measures relationships and not just references, giving more value to relationships.

Table 2. Top 10 authors by H-index (Nco-author = Number of coauthors, Ncite = Number of cites, Ndoc = Number of documents).

Author Scopus ID	Name	H-Index	Nco-Author	Ncite	Ndoc	City	Country	Affiliation
35463345800	Gratzel M.	230	2222	258107	1471	Lausanne	Switzerland	Swiss Federal Institute of Technology, Lausanne
55711979600	Whitesides G.	187	1268	159450	1008	Cambridge	United States	Harvard University
7403027697	Xia Y.	186	952	136496	779	Atlanta	United States	Georgia Institute of Technology
7103185149	Antonietti M.	156	1047	84786	789	Golm	Germany	Max Planck Institut fur Kolloid Und Grenzflächenforschung Potsdam
35207974600	Cui Y.	155	994	99093	532	Menlo Park	United States	Stanford Linear Accelerator Center
7403931988	Yang P.	150	706	102859	408	Berkeley	United States	University of California, Berkeley
56605567400	Alivisatos A.	143	934	108954	445	Berkeley	United States	University of California, Berkeley
7403489871	Zhao D.	138	1189	77076	685	Shanghai	China	Fudan University
35463772200	Nazeeruddin M.	133	1090	82713	576	Lausanne	Switzerland	Swiss Federal Institute of Technology, Lausanne
56422845100	Jiang L.	130	2113	72789	1256	Beijing	China	Technical Institute of Physics and Chemistry Chinese Academy of Sciences

Figure 11. Location of the Author with the highest H-index with research in Solar Coating.

If we analyze the top 10 of the authors considered in the analyzed communities (Table 3), it is observed that they have lower H-index than those shown in Table 2 and, in this case, if they correspond to the affiliations in Table 1. An author from Belgium appears and another from France, which does not appear from its affiliations in Table 2, despite being in positions 5 and 8 of Table 3.

Table 3. Most important 10 for H-index analyzed in the communities (Nco-author = Number of coauthors, Ncite = Number of cites, Ndoc = Number of documents).

Author Scopus ID	Name	H-Index	Nco-Author	Ncite	Ndoc	City	Country	Affiliation
6701805412	Ballif C.	65	750	13344	461	Lausanne	Switzerland	Swiss Federal Institute of Technology, Lausanne
56216991600	Rech B.	48	514	10525	321	Berlin	Germany	Helmholtz-Zentrum Berlin für Materialien und Energie (HZB)
6603242760	Blasi B.	21	201	1542	105	Freiburg im Breisgau	Germany	Fraunhofer Institute for Solar Energy Systems ISE
7004315658	Haug F.	37	282	5174	171	Lausanne	Switzerland	Swiss Federal Institute of Technology, Lausanne
56597035200	Poortmans J.	44	781	8334	523	Leuven	Belgium	Interuniversity Micro-Electronics Center at Leuven
55505896100	Zhao Y.	21	338	2607	462	Tianjin	China	Nankai University
55931076700	Zhang X.	17	420	2020	383	Tianjin	China	Nankai University
6602741595	Roca i Cabarrocas P.	43	683	7634	497	Palaiseau	France	Laboratoire de Physique des Interfaces et des Couches Minces
7402736621	Yi J.	33	617	4594	471	Jongno-gu	South Korea	Sungkyunkwan University
7006823424	Atwater H.	89	726	41733	654	Pasadena	United States	California Institute of Technology

4. Discussion

The scientific contributions that were published in the Solar Coating field from 1964 to June 2019 make a total of 6440 and 127 new ones published as of September. Until 1982, the publications were sporadic with a couple of publications per year, and then increased in 1990, but did not exceed 20 annual publications. From then on, the contributions are more numerous until 2004, being marked with 97 publications. From 2005 to 2019, this period is marked with a high concentration of publications (83%), due to the "renewable boom". The fact that the highest concentration has occurred, since 2005 might be due to global awareness of environmental concerns, such as global warming and the greenhouse effect, as shown by the agreement in Kyoto, Japan, on 11 December 1997 and entered into force on February 16, 2005. This has meant that the scientific community has turned to providing solutions; one of the clean energies that is most committed to the future is photovoltaic solar energy [73]. The main forms of publications are in original articles with 62.6% and in conferences with 32.6%, followed by far by review articles and book chapters with 1.9% and 1.1%, respectively. Book contributions are scarce, with only six published < 0.01% [74–80] and all with little impact on the number of references below 22. In [16], the authors analyze the bibliometric in energy saving and obtain a percentage of publication of articles of 50.7%, conferences 43.1%, and in books a percentage lower than 1%, which is very much in line with those that were obtained in this work.

English is the most used language for publications with 96.8% of the total publications, followed by Chinese, with a percentage of 2.6%, and the rest of the languages have a representation lower than < 1% (Russian, Japanese, Korean, German, Spanish, French, Lithuanian, Finnish, and Malay). It is usual for the English language to be the most widespread in scientific publications due to the edition standards of the journals to have maximum dissemination, regardless of the country of origin of the headquarters of the publisher. Only China produces a number of scientific publications in its own language, such as Taiyangneng Xuebao/Acta Energiae Solaris Sinica, Cailiao Gongcheng/Journal of Materials Engineering among others. The demand for energy increases with population growth and China is one of the largest consumers, due to its industrial expansion and population growth, which motivates China to investigate the production of clean energy, as seen in the results shown in Tables 1–3.

The thematic areas that are most used for scientific dissemination are mainly Physics and Astronomy (25.9%), Materials Science (23.6%), Engineering (21.6%), Energy 6.4%), Computer Science (5.6%), Chemistry, Mathematics (4.6%), Chemical Engineering (4.6%), and the rest with a contribution of less than 3%. The results show the great involvement of researchers from different disciplines, such as engineering, materials sciences, chemistry, and mathematics, among others, having all the purpose of contributing to improvements in solar energy collection. The contributions of the application of nanomaterials and nanostructures to the collection surfaces have allowed for improving efficiency by trapping light and generating new flexible, bifacial panels that have expanded the use of these panels as an architectural tool in industrial buildings and facilities, electric locomotion, and solar powered devices, thereby reducing the use of fossil fuels and CO_2 emissions.

5. Conclusions

This work has revealed data related to Solar Coating from 1957 to 2019. The total contributions found reached 6440 documents with a total of 21,301 relations between authors, where only 3924 with 20,849 relations after the use of bibliometric techniques. The study of efficiency improvements has driven a large number of contributions in different sub-themes, where each community develops its research on different perspectives, although they all have the improvement of light trapping as a system for improving efficiency in common. Publications have focused on six categories or communities Material Solar Cell, Thin Film and Polycrystalline, Organic solar cells Thin film a-Si: H, Optical design, and Light trapping. The communities with the highest representation are the most generic: The Material Solar Cell (42%), Thin film a-Si: H (25%), and Light trapping (11%). However, from a different perspective study aspect, they are provided with a main nucleus and multiple relations with the rest of the communities that nourish them of references. Organic solar cells (8.3%), Optical design (8%), and Thin Film and Polycrystalline (6%) are the other three communities. In addition to having less representativeness, these communities do not have a main nucleus, but small nuclei with a multitude of relationships in and between communities, with organic solar cells having 19 major nuclei, although much lower than all of the main ones of the more generic communities. The most repeated keywords correspond to the generic communities, which is consistent because they are the ones with the greatest ties and the most referenced publications. There are communities, like Optical design with keywords, which are very specific to their theme. This is because this community is tangential to Solar Coating incorporating research on the theme of Optics that are not specifically developed for solar cells, but their advances substantially influence the improvements of light capture, which is an essential theme for Solar Coating.

In total, there are 14,849 authors who research in 26 by subject area, China (3345 researchers) and the United States (2634 researchers) are from the countries with the greatest contributions in scientific developments in all matters related to energy, also in energy saving. The worldwide distribution is established between Asia (38%), Europe (16%), and America, with only the United States (17.7%), and there is no significant representation of the African continent. If we consider the authors' affiliation with a top 20, 12 of the 20 belong to the Asian continent (65%). It is interesting how Switzerland, which did not appear among the 10 countries with the highest productivity in Solar Coating, is the second research institution in terms of the number of contributions with 6.6% of the top 20. In Europe, Germany stands out with the Fraunhofer Institute for Solar Energy Systems centers ISE and Helmholtz-Zentrum Berlin für Materialien und Energie (HZB) (11.1%). In the USA, which has 18% of the worldwide publications, with respect to the ranking of the top 20 affiliations only three research institutions are present, from them the best positioned is Stanford University.

The analysis of the 10 authors with the highest H-index in Scopus in relation to the search performed does not correspond to the order by institution, for example, the author with the highest H index of 230, 1471 published documents, and 258,107 citations belongs to the Swiss Federal affiliation Institute of Technology of Lausanne, (Switzerland). On the other hand, the first author in the Top 20 by affiliation is from National Taiwan University (Taiwan).

The main forms of publications are in original articles (62.6%), conferences (32.6%), followed by review articles (1.9%) and book chapters (1.1%), with scarce book contributions (<0.01%). As for the language, English (96.8%) is the most used for disseminating publications due to the standards of edition of the journals to have maximum dissemination, regardless of the country of origin of the headquarters of the publisher. The second language used is Chinese (2.6%), as it produces a significant number of scientific publications with international repercussions in their own language. All other languages have a representation of less than <1%.

The most used thematic areas are very diverse; this shows the great involvement of researchers from different disciplines, such as engineering, materials sciences, chemistry and mathematics, among others, all with the purpose of contributing to improvements in solar energy collection.

The final conclusion of the present work shows that the research on this subject is not completed and it requires even more research with different considerations that improve solar cell efficiencies. The joint collaboration of all researchers is required, with their valuable contributions opening new perspectives regarding the necessary improvement due to the growing need for global demand for energy consumption and the awareness of consumers and political entities of environmental care and implementation of renewable energy.

Author Contributions: N.N. formed the manuscript; A.A. and F.M.-A. developed the figures and tables; A.A. and D.E.K. contributed to the search of date and the realized of the maps; F.M.-A. and N.N. redacted the paper. D.E.K. checked the whole manuscript.

Funding: Under I+D+I Project—FEDER Andalucía 2014-2020 Operational Programme, UAL18-TIC-A025-A.

Acknowledgments: The Ministry of Economic and Competitiveness of Spain financed this work, under Project TEC2014-60132-P, in part by Innovation, Science and Enterprise, Andalusian Regional Government through the Electronics, Communications and Telemedicine TIC019 Research Group of the University of Almeria, Spain and in part by the European Union FEDER Program and CIAMBITAL Group.

Conflicts of Interest: The authors declare no conflict of interest.

References

1. McCollum, D.; Gomez Echeverri, L.; Riahi, K.; Parkinson, S. Affordable and clean energy: Ensure access to affordable, reliable, sustainable, and modern energy for all. In *Atlas of Sustainable Development Goals 2018: From World Development Indicators*; International Bank for Reconstruction and Development/The World Bank: Washington, WA, USA, 2018; pp. 26–29.
2. Hansen, K.; Vad Mathiesen, B. Comprehensive assessment of the role and potential for solar thermal in future energy systems. *Sol. Energy* **2018**, *169*, 144–152. [CrossRef]
3. Jia, T.; Dai, Y.; Wang, R. Refining energy sources in winemaking industry by using solar energy as alternatives for fossil fuels: A review and perspective. *Renew. Sustain. Energy Rev.* **2018**, *88*, 278–296. [CrossRef]
4. Rogelj, J.; Den Elzen, M.; Höhne, N.; Fransen, T.; Fekete, H.; Winkler, H.; Meinshausen, M. Paris Agreement climate proposals need a boost to keep warming well below 2 °C. *Nature* **2016**, *534*, 631–639. [CrossRef] [PubMed]
5. Kilinc-Ata, N. The evaluation of renewable energy policies across EU countries and US states: An econometric approach. *Energy Sustain. Dev.* **2016**, *31*, 83–90. [CrossRef]
6. Jiaru, H.; Xiangzhao, F. An evaluation of China's carbon emission reduction policies on urban transport system. *J. Sustain. Dev. Law Policy* **2016**, *6*, 31. [CrossRef]
7. Vasiliev, M.; Nur-e-alam, M.; Alameh, K. Initial field testing results from building-integrated solar energy harvesting windows installation in Perth, Australia. *Appl. Sci.* **2019**, *9*, 4002. [CrossRef]
8. Catchpole, K.R.; Mokkapati, S.; Beck, F.; Wang, E.C.; McKinley, A.; Basch, A.; Lee, J. Plasmonics and nanophotonics for photovoltaics. *MRS Bull.* **2011**, *36*, 461–467. [CrossRef]
9. Ouyang, Z.; Pillai, S.; Beck, F.; Kunz, O.; Varlamov, S.; Catchpole, K.R.; Green, M.A. Effective light trapping in polycrystalline silicon thin-film solar cells by means of rear localized surface plasmons. *Appl. Phys. Lett.* **2010**, *96*, 261109. [CrossRef]
10. Rand, J.A.; Basore, P.A. Light-trapping silicon solar cells-experimental results and analysis. In Proceedings of the Conference Record of the IEEE Photovoltaic Specialists Conference, Las Vegas, NV, USA, 7–11 October 1991; pp. 192–197.
11. Bokor, N.; Davidson, N. Curved diffractive optical elements: Design and applications. *Prog. Opt.* **2005**, *48*, 107–148.
12. Cirino, G.A.; Verdonck, P.; Mansano, R.D.; Pizolato, J.C., Jr.; Mazulquim, D.B.; Neto, L.G. Digital holography: Computer-generated holograms and diffractive optics in scalar diffraction domain. In *Holography-Different Fields of Application*; Monroy, F., Ed.; IntechOpen: London, UK, 2011. [CrossRef]

13. Luo, M.; Zhou, L.; Yuan, J.; Zhu, C.; Cai, F.; Hai, J.; Zou, Y. A new non-fullerene acceptor based on the heptacyclic benzotriazole unit for efficient organic solar cells. *J. Energy Chem.* **2020**, *42*, 169–173. [CrossRef]
14. Montoya, F.G.; Alcayde, A.; Baños, R.; Manzano-agugliaro, F. Telematics and Informatics A fast method for identifying worldwide scientific collaborations using the Scopus database. *Telemat. Inf.* **2018**, *35*, 168–185. [CrossRef]
15. Mongeon, P.; Paul-Hus, A. The journal coverage of Web of Science and Scopus: A comparative analysis. *Scientometrics* **2016**, *106*, 213–228. [CrossRef]
16. Cruz-Lovera, C.; Perea-Moreno, A.-J.; de la Cruz-Fernández, J.L.; Montoya, F.G.; Alcayde, A.; Manzano-Agugliaro, F. Analysis of Research Topics and Scientific Collaborations in Energy Saving Using Bibliometric Techniques and Community Detection. *Energies* **2019**, *12*, 2030. [CrossRef]
17. Jacomy, M.; Venturini, T.; Heymann, S.; Bastian, M. ForceAtlas2, a continuous graph layout algorithm for handy network visualization designed for the Gephi software. *PLoS ONE* **2014**, *9*, e98679. [CrossRef] [PubMed]
18. Garnett, E.; Yang, P. Light trapping in silicon nanowire solar cells. *Nano Lett.* **2010**, *10*, 1082–1087. [CrossRef] [PubMed]
19. Yu, Z.; Raman, A.; Fan, S. Fundamental limit of nanophotonic light trapping in solar cells. *Next Gener. Photonic Cell Technol. Sol. Energy Convers.* **2010**, *7772*, 77720Z.
20. Bermel, P.; Luo, C.; Zeng, L.; Kimerling, L.C.; Joannopoulos, J.D. Improving thin-film crystalline silicon solar cell efficiencies with photonic crystals. *Opt. Express* **2007**, *15*, 16986–17000. [CrossRef]
21. Kelzenberg, M.D.; Boettcher, S.W.; Petykiewicz, J.A.; Turner-Evans, D.B.; Putnam, M.C.; Warren, E.L.; Atwater, H.A. Enhanced absorption and carrier collection in Si wire arrays for photovoltaic applications. *Nat. Mater.* **2010**, *9*, 239–244. [CrossRef]
22. Mokkapati, S.; Catchpole, K.R. Nanophotonic light trapping in solar cells. *J. Appl. Phys.* **2012**, *112*, 10. [CrossRef]
23. Brongersma, M.L.; Cui, Y.; Fan, S. Light management for photovoltaics using high-index nanostructures. *Nat. Mater.* **2014**, *13*, 451–460. [CrossRef]
24. Yu, Z.; Raman, A.; Fan, S. Fundamental limit of light trapping in grating structures. *Opt. Express* **2010**, *18*, A366–A380. [CrossRef] [PubMed]
25. Yang, W.; Yu, H.; Wang, Y. Light transmission through periodic sub-wavelength textured surface. *J. Opt.* **2013**, *15*, 5. [CrossRef]
26. Wang, K.X.; Yu, Z.; Liu, V.; Cui, Y.; Fan, S. Absorption enhancement in ultrathin crystalline silicon solar cells with antireflection and light-trapping nanocone gratings. *Nano Lett.* **2012**, *12*, 1616–1619. [CrossRef]
27. Aberle, A.G.; Widenborg, P.I. Crystalline Silicon Thin-Film Solar Cells Via High-Temperature and Intermediate-Temperature Approaches. In *Handbook of Photovoltaic Science and Engineering*, 2nd ed.; Wiley: New York, NY, USA, 2011.
28. Eisenhauer, D.; Trinh, C.T.; Amkreutz, D.; Becker, C. Light management in crystalline silicon thin-film solar cells with imprint-textured glass superstrate. *Sol. Energy Mater. Sol. Cells* **2019**, *200*, 109928. [CrossRef]
29. Xue, C.; Rao, J.; Varlamov, S. A novel silicon nanostructure with effective light trapping for polycrystalline silicon thin film solar cells by means of metal-assisted wet chemical etching. *Phys. Stat. Solidi Appl. Mater. Sci.* **2013**, *210*, 2588–2591. [CrossRef]
30. Gall, S.; Becker, C.; Conrad, E.; Dogan, P.; Fenske, F.; Gorka, B.; Rech, B. Polycrystalline silicon thin-film solar cells on glass. *Sol. Energy Mater. Sol. Cells* **2009**, *93*, 1004–1008. [CrossRef]
31. Mehmood, H.; Tauqeer, T.; Hussain, S. Recent progress in silicon-based solid-state solar cells. *Int. J. Electron.* **2018**, *105*, 1568–1582. [CrossRef]
32. Jia, G.; Andrä, G.; Gawlik, A.; Schönherr, S.; Plentz, J.; Eisenhawer, B.; Falk, F. Nanotechnology enhanced solar cells prepared on laser-crystallized polycrystalline thin films (<10 μm). *Sol. Energy Mater. Sol. Cells* **2014**, *126*, 62–67.
33. Xiao, S.; Wang, T.; Liu, Y.; Xu, C.; Han, X.; Yan, X. Tunable Light Trapping and Absorption Enhancement with Graphene Ring Arrays. *Phys. Chem. Chem. Phys.* **2016**, *18*, 26661–26669. [CrossRef]
34. Abenante, L. Optical path length factor at near-bandgap wavelengths in Si solar cells. *IEEE Trans. Electron Devices* **2006**, *53*, 3047–3053. [CrossRef]
35. Yablonovitch, E.; Cody, G.D. Intensity Enhancement in Textured Optical Sheets for Solar Cells. *IEEE Trans. Electron Devices* **1982**, *29*, 300–305. [CrossRef]

36. Becker, C.; Amkreutz, D.; Sontheimer, T.; Preidel, V.; Lockau, D.; Haschke, J.; Jogschies, L.; Klimm, C.; Merkel, J.J.; Plocica, P.; et al. Polycrystalline silicon thin-film solar cells: Status and perspectives. *Sol. Energy Mater. Sol. Cells* **2013**, *119*, 112–123. [CrossRef]

37. Dross, F.; Baert, K.; Bearda, T.; Deckers, J.; Depauw, V.; El Daif, O.; Gordon, I.; Gougam, A.; Govaerts, J.; Granata, S.; et al. Crystalline thin-foil silicon solar cells: Where crystalline quality meets thin-film processing. *Prog. Photovolt. Res. Appl.* **2012**, *20*, 770–784. [CrossRef]

38. Becker, C.; Preidel, V.; Amkreutz, D.; Haschke, J.; Rech, B. Double-side textured liquid phase crystallized silicon thin-film solar cells on imprinted glass. *Sol. Energy Mater. Sol. Cells* **2015**, *135*, 2–7. [CrossRef]

39. Jeong, E.; Zhao, G.; Song, M.; Yu, S.M.; Rha, J.; Shin, J.; Cho, Y.-R.; Yun, J. Simultaneous improvements in self-cleaning and light-trapping abilities of polymer substrates for flexible organic solar cells. *J. Mater. Chem. A* **2018**, *6*, 2379–2387. [CrossRef]

40. Tang, Z.; Tress, W.; Inganäs, O. Light trapping in thin film organic solar cells. *Mater. Today* **2014**, *17*, 389–396. [CrossRef]

41. Ko, D.H.; Tumbleston, J.R.; Gadisa, A.; Aryal, M.; Liu, Y.; Lopez, R.; Samulski, E.T. Light-trapping nano-structures in organic photovoltaic cells. *J. Mater. Chem.* **2011**, *21*, 16293–16303. [CrossRef]

42. Li, X.; Choy, W.C.H.; Lu, V.; Sha, W.E.I.; Ho, A.H.P. Efficiency enhancement of organic solar cells by using shape-dependent broadband plasmonic absorption in metallic nanoparticles. *Adv. Funct. Mater.* **2013**, *23*, 2728–2735. [CrossRef]

43. Feng, L.; Niu, M.; Wen, Z.; Hao, X. Recent advances of plasmonic organic solar cells: Photophysical investigations. *Polymers* **2018**, *10*, 123. [CrossRef]

44. Tvingstedt, K.; Dal Zilio, S.; Inganäs, O.; Tormen, M. Trapping light with micro lenses in thin film organic photovoltaic cells. *Opt. Express* **2008**, *16*, 21608–21615. [CrossRef]

45. Xiao, X.; Zhang, Z.; Xie, S.; Liu, Y.; Hu, D.; Du, J. Enhancing light harvesting of organic solar cells by using hybrid microlenses. *Opt. Appl.* **2015**, *45*, 89–100.

46. Rim, S.B.; Zhao, S.; Scully, S.R.; McGehee, M.D.; Peumans, P. An effective light trapping configuration for thin-film solar cells. *Appl. Phys. Lett.* **2007**, *91*, 10–13. [CrossRef]

47. Müller-Meskamp, L.; Kim, Y.H.; Roch, T.; Hofmann, S.; Scholz, R.; Eckardt, S.; Lasagni, A.F. Efficiency enhancement of organic solar cells by fabricating periodic surface textures using direct laser interference patterning. *Adv. Mater.* **2012**, *24*, 906–910. [CrossRef] [PubMed]

48. Jeon, I.; Yoon, J.; Kim, U.; Lee, C.; Xiang, R.; Shawky, A.; Xi, J.; Byeon, J.; Lee, H.M.; Choi, M.; et al. High-Performance Solution-Processed Double-Walled Carbon Nanotube Transparent Electrode for Perovskite Solar Cells. *Adv. Energy Mater.* **2019**, *9*, 1901204. [CrossRef]

49. Ahn, N.; Son, D.Y.; Jang, I.H.; Kang, S.M.; Choi, M.; Park, N.G. Highly reproducible perovskite solar cells with average efficiency of 18.3% and best efficiency of 19.7% fabricated via Lewis base adduct of lead (II) iodide. *J. Am. Chem. Soc.* **2015**, *137*, 8696–8699. [CrossRef]

50. Gan, Q.; Bartoli, F.J.; Kafafi, Z.H. Plasmonic-enhanced organic photovoltaics: Breaking the 10% efficiency barrier. *Adv. Mater.* **2013**, *25*, 2385–2396. [CrossRef]

51. Ou, Q.D.; Li, Y.Q.; Tang, J.X. Light manipulation in organic photovoltaics. *Adv. Sci.* **2016**, *3*, 1600123. [CrossRef]

52. Müller, J.; Rech, B.; Springer, J.; Vanecek, M. TCO and light trapping in silicon thin film solar cells. *Sol. Energy* **2004**, *77*, 917–930. [CrossRef]

53. Battaglia, C.; Hsu, C.M.; Söderström, K.; Escarré, J.; Haug, F.J.; Charrière, M.; Cui, Y. Light trapping in solar cells: Can periodic beat random? *ACS Nano* **2012**, *6*, 2790–2797. [CrossRef]

54. Ferry, V.E.; Verschuuren, M.A.; Li, H.B.; Verhagen, E.; Walters, R.J.; Schropp, R.E.; Polman, A. Light trapping in ultrathin plasmonic solar cells. *Opt. Express* **2010**, *18*, 237–245. [CrossRef]

55. Shah, A.V.; Schade, H.; Vanecek, M.; Meier, J.; Vallat-Sauvain, E.; Wyrsch, N.; Bailat, J. Thin-film silicon solar cell technology. *Prog. Photovolt. Res. Appl.* **2004**, *12*, 113–142. [CrossRef]

56. He, Y.; Chen, M.; Zhou, J.; Peng, T.; Ren, Z. Applying Light Trapping Structure to Solar Cells: An Overview. *Cailiao Daobao* **2018**, *32*, 696–707.

57. Berginski, M.; Hüpkes, J.; Schulte, M.; Schöpe, G.; Stiebig, H.; Rech, B.; Wuttig, M. The effect of front ZnO:Al surface texture and optical transparency on efficient light trapping in silicon thin-film solar cells. *J. Appl. Phys.* **2007**, *101*, 7. [CrossRef]

58. Tan, H.; Santbergen, R.; Smets, A.H.M.; Zeman, M. Plasmonic light trapping in thin-film silicon solar cells with improved self-assembled silver nanoparticles. *Nano Lett.* **2012**, *12*, 4070–4076. [CrossRef]

59. Ferry, V.E.; Verschuuren, M.A.; Lare, M.C.V.; Schropp, R.E.; Atwater, H.A.; Polman, A. Optimized Spatial Correlations for Broadband Light Trapping. *Nano Lett.* **2011**, *11*, 4239–4245. [CrossRef]

60. Zhu, J.; Hsu, C.M.; Yu, Z.; Fan, S.; Cui, Y. Nanodome solar cells with efficient light management and self-cleaning. *Nano Lett.* **2010**, *10*, 1979–1984. [CrossRef]

61. Herkommer, A.M.; Reichle, R.; Häfner, M.; Pruss, C. Design and simulation of diffractive optical components in fast optical imaging systems. *Opt. Des. Eng. IV* **2011**, *8167*, 816708.

62. Wood, A.P. Design of infrared hybrid refractive–diffractive lenses. *Appl. Opt.* **1992**, *31*, 2253. [CrossRef]

63. Feng, L.; Xiping, X.; Xiangyang, S. Design of infrared (IR) hybrid refractive/diffractive lenses for target detecting/tracking. *Acta Opt. Sin.* **2010**, *30*, 2084–2088. [CrossRef]

64. Škereň, M.; Svoboda, J.; Květoň, M.; Hopp, J.; Possolt, M.; Fiala, P. Diffractive elements for correction of chromatic aberrations of illumination systems. *EPJ Web Conf.* **2013**, *48*, 00025. [CrossRef]

65. Flores, A.; Wang, M.R.; Yang, J.J. Achromatic hybrid refractive-diffractive lens with extended depth of focus. *Appl. Opt.* **2004**, *43*, 5618. [CrossRef]

66. Wyrowski, F.; Bryngdahl, O. Digital holography as part of diffractive optics. *Rep. Prog. Phys.* **1991**, *54*, 1481. [CrossRef]

67. Mait, J.N. Understanding diffractive optic design in the scalar domain. *JOSA A* **1995**, *12*, 2145–2158. [CrossRef]

68. Bryngdahl, O.; Wyrowski, F. I Digital Holography–Computer-Generated Holograms. *Prog. Opt.* **1990**, *28*, 1–86.

69. Pillai, S.; Catchpole, K.R.; Trupke, T.; Green, M.A. Surface plasmon enhanced silicon solar cells. *J. Appl. Phys.* **2007**, *101*, 9. [CrossRef]

70. Catchpole, K.R.; Polman, A. Design principles for particle plasmon enhanced solar cells. *Appl. Phys. Lett.* **2008**, *93*, 10–13. [CrossRef]

71. Beck, F.J.; Polman, A.; Catchpole, K.R. Tunable light trapping for solar cells using localized surface plasmons. *J. Appl. Phys.* **2009**, *105*, 114310. [CrossRef]

72. Mokkapati, S.; Beck, F.J.; Polman, A.; Catchpole, K.R. Designing periodic arrays of metal nanoparticles for light-trapping applications in solar cells. *Appl. Phys. Lett.* **2009**, *95*, 053115. [CrossRef]

73. Yang, Y.; Pillai, S.; Mehrvarz, H.; Kampwerth, H.; Ho-Baillie, A.; Green, M.A. Enhanced light trapping for high efficiency crystalline solar cells by the application of rear surface plasmons. *Sol. Energy Mater. Sol. Cells* **2012**, *101*, 217–226. [CrossRef]

74. Deetjen, T.A.; Conger, J.P.; Leibowicz, B.D.; Webber, M.E. Review of climate action plans in 29 major U.S. cities: Comparing current policies to research recommendations. *Sustain. Cities Soc.* **2018**, *41*, 711–727. [CrossRef]

75. Boriskina, S.; Zheludev, N.I. (Eds.) *Singular and Chiral Nanoplasmonics*; CRC Press: New York, NY, USA, 2014.

76. Gangopadhyay, U.; Dutta, S.K.; Saha, H. *Texturization and Light Trapping in Silicon Solar Cells*; Nova Science Publishers Inc: New York, NY, USA, 2009; pp. 44–62.

77. Kane, D.; Micolich, A.; Rabeau, J. *Nanotechnology in Australia: Showcase of Early Career Research*; Pan Stanford Publishing Pte. Ltd.: Singapore, 2011; p. 463.

78. Clarke, J. *Energy Simulation in Building Design*; Routledge: London, UK, 2001. [CrossRef]

79. Fonash, S.J. *Introduction to Light Trapping in Solar Cell and Photo-Detector Devices*; Elsevier: Amsterdam, The Netherlands, 2015.

80. Tiwari, A.; Boukherroub, R.; Sharon, M. *Solar Cell Nanotechnology*; Wiley-Scrivener: Austin, TX, USA, 2013.

 *coatings*

Article

Advanced Analysis of Corroded Solar Reflectors

Francisco Buendía-Martínez [1], Aránzazu Fernández-García [1,*], Florian Sutter [2], Loreto Valenzuela [1] and Alejandro García-Segura [1]

[1]   CIEMAT-Plataforma Solar de Almería, Ctra. Senés, 04200 Tabernas, Spain; francisco.buendia@psa.es (F.B.-M.); loreto.valenzuela@psa.es (L.V.); alexgarsek@gmail.com (A.G.-S.)

[2]   German Aerospace Center (DLR), Ctra. Senés, 04200 Tabernas, Spain; florian.sutter@dlr.de

*   Correspondence: arantxa.fernandez@psa.es; Tel.: +34-950-387-950

Received: 26 September 2019; Accepted: 8 November 2019; Published: 11 November 2019

Abstract: The corrosion of the reflective layer is one of the main degradation mechanisms of solar reflectors. However, the appropriate assessment of the corroded reflector samples is not accomplished by the current analysis techniques. On the one hand, the reflectance measurement protocol of non-damaged solar reflectors for concentrating solar thermal technologies is widely addressed in the SolarPACES reflectance guideline. However, this methodology is not adequate for reflectors whose surface is partially corroded by many kind of corrosion agents. In this work, a new measurement technique to properly assess corroded samples was developed. To check the usefulness of the method, several damaged samples (subjected to two accelerated aging tests) were evaluated with the conventional technique and with the improved one. The results showed that a significant discrepancy is observed between the two methods for heavily corroded samples, with average reflectance differences of 0.053 ppt. The visualization of the reflector images illustrated that the improved method is more reliable. On the other hand, both the corrosion products formed and the corrosion rates were identified after each corrosive test. The chemical atmosphere significantly affects the products formed, whereas the corrosion rates are influenced by the test conditions and the reflector quality.

Keywords: concentrating solar thermal energy; corroded solar reflector; improved measurement method; corrosion product; corrosion rate; monochromatic specular reflectance; solar hemispherical reflectance

---

1. Introduction

The increase of the greenhouse gases in recent years, especially the CO_2 emissions [1], has led to a change in the energy mix [2,3]. The traditional energy sources such as coal, oil, and natural gas are being replaced by other sources with a less aggressive impact on the environment. This change is led by renewables energies [4,5]. Within them the importance of the solar energy should be noted as the best alternative to mitigate the effects originated by the fossil sources because its repercussion on the environment is almost negligible [6–8]. For this reason, solar energy has experienced the renewable energies supply highest average annual growth rate in the world (56.9%), from 1990 to 2015 [9]. Solar energy can be classified into solar thermal energy (STE) or photovoltaic (PV) energy, depending on the energy conversion process. In addition, STE can be divided, depending on the concentration, into concentrating solar thermal (CST) or non-concentrating solar thermal technologies. Regarding CST energy, the total capacity installed worldwide is 5.5 GW, Spain being the country with the highest contribution, 2.3 GW, i.e., the 42% of the total capacity installed [10,11]. According to the International Energy Agency's (IEA) forecast for 2050 [11], an 11% of the worldwide energy mix will be provided by CST systems.

Reflectors, commonly called mirrors, are a crucial component in CST technologies whose goal is to concentrate the radiation in order to transform solar energy into thermal energy [12]. Depending on the reflective layer, mirrors can be classified into aluminum or silvered reflectors [13]. To achieve suitable operating conditions in CST technologies, a reflector with high efficiency should be installed to reach the maximum plant's output [14].

The optical parameter that correctly quantifies the efficiency of a solar reflector is the reflectance, ρ. The measurement of this parameter is a non-trivial issue because many variables are involved in the reflection process, such as the wavelength, λ, the beam divergence of the incident light source, φ_i, and the incidence angle, θ_i of the incoming solar beam, as well as the acceptance angle, φ, of the receiver or measurement device detector. A group of experts have been working on the proper definition of the reflectance under the framework of SolarPACES Task III, within the IEA. Several agreements reached by this group are collected in a document in which current version is titled "Parameters and method to evaluate reflectance properties of reflector materials for concentrating solar power technology" (hereinafter "SolarPACES Reflectance Guideline") [15]. This document has been used as a reference by the CST community for the last years and it has even been mentioned by the first standard published about solar reflectors' durability testing, the UNE 206016:2018 [16], as the measurement method to assess durability experiments.

The measurement protocol described in this guideline, in order to measure new and clean solar reflectors, has proven to be very accurate and easy-to-use for any laboratory or company equipped with conventional commercial instruments [17,18]. However, the proposed conventional method used to characterize aged and/or soiled reflector samples is insufficient and inaccurate since only some specific spots of the samples are assessed and it does not take into consideration θ_i and φ dependence. Recently, several research works which focused on the characterization of soiled reflectors were published [19–22]. However, the evaluation of aged mirrors is a challenging topic that has only been addressed up to now with a prototype instrument specially developed by DLR for that purpose [23].

The durability of the solar reflectors is one of the most important parameters to consider when a CSP plant is designed [13]. In order to guarantee that the reflector is suitable to maintain its optical properties during the whole lifetime of the plant, several aging tests should be performed to assure the reliability of the material. For this purpose, corrosion tests must be carried out to simulate both the corrosion conditions provoked by salty climates, where the concentration of Cl^- ions are higher than usual, as well as polluted atmosphere originated by industries, where the concentration of harmful gases (such as SO_2, NO_2, or H_2S) is extremely high. These real outdoor conditions significantly affect the lifetime of the solar materials because the reflector layer (normally silver) reacts with the chemical compounds of corrosive environments. Consequently, the analysis of the corrosion parameters is a crucial aspect to be considered in the durability studies.

This paper presents an advanced method to analyze corroded solar reflectors. On the one hand, an improved optical measurement technique that accurately measures aged mirrors degraded due to corrosion mechanisms was developed. This method is based on the use of photographic images, the two commercial instruments normally employed in the conventional method (that is, reflectometers and spectrophotometers), and the application of an innovative measurement protocol. Results obtained from this new technique have highly improved the understanding of the overall efficiency decrease produced along the whole solar reflector's surface due to the corrosion effects. On the other hand, the corrosion products appearing on the samples were identified and related to the chemicals added to each corrosive test, and the influence of the materials quality and the testing conditions were related with both the corrosion rates and the total corroded areas.

2. Materials and Methods

This section includes the description of the optical reflectance parameters normally used to characterize solar reflectors, the reflector materials measured in this study, the durability tests applied to these reflectors, and the measurement equipment employed. Finally, both the conventional optical measurement technique and the improved technique proposed in this work are presented.

2.1. Reflectance Definitions

The optical parameter to correctly characterize a solar reflector is the solar-weighted near-specular reflectance, $\rho_{s,\varphi}([\lambda_a, \lambda_b],\theta_i,\varphi)$, which is defined as ratio of the radiant flux reflected from a surface in the specular direction (and collected into φ) to that of the incident radiation flux (coming with an angle θ_i and weighted in the range from λ_a to λ_b) [14]. Unfortunately, the employed devices to measure it are lab-prototypes [24–27], which are not commercially available yet. Consequently, solar reflectors are normally assessed by an indirect method based on the combination of the following reflectance values:

- Solar-weighted hemispherical reflectance, $\rho_{s,h}([\lambda_a, \lambda_b],\theta_i,h)$, which is calculated by weighting the hemispherical reflectance spectrum, $\rho_{\lambda,h}$, with the solar direct irradiance, G_b, on the earth surface for each λ_i, according to Equation (1) [28].

$$\rho_{s,h}([\lambda_a,\lambda_b],\theta_i,h) = \frac{\int_{\lambda_a}^{\lambda_b} \rho_{\lambda,h}(\lambda,\theta_i,h)G_b(\lambda)\mathrm{d}\lambda}{\int_{\lambda_a}^{\lambda_b} G_b(\lambda)\mathrm{d}\lambda} \tag{1}$$

where $\rho_{\lambda,h}$ is the ratio of the incident and emitted energy flux of a surface within the complete hemisphere [14], measured with a spectrophotometer. For European and North American latitudes, typical solar direct irradiance spectra are given by the current standard ASTM G173-03 (air mass AM 1.5) [29].

- Monochromatic near-specular reflectance, $\rho_{\lambda,\varphi}(\lambda,\theta_i,\varphi)$, which is the ratio of incident and emitted energy flux of a surface in the specular direction [14]. It is measured with a reflectometer.
- Monochromatic hemispherical reflectance, $\rho_{\lambda,h}(\lambda,\theta_i,h)$, which is the value of the spectral hemispherical reflectance at the same λ of the $\rho_{\lambda,\varphi}$ measured. It is used to calculate the specularity of the reflectors, that is, the ratio between $\rho_{\lambda,\varphi}$ and $\rho_{\lambda,h}$.

2.2. Materials

Second surface silvered-glass reflectors are the most commonly used materials for CST technologies [30,31]. Consequently, this work is focused on the analysis of this type of solar reflectors. They were composed of a low-iron glass substrate (1–4 mm thickness) coated with a silver reflective layer on the backside (see Figure 1). To protect the silver on the backside, the mirror backing system consists of a copper layer and several protective paints. All the reflector samples analyzed were 100 mm by 100 mm and they featured an original edge, in which the metal layers are completely covered and protected by paint layers (protected edge), and three fractured (or unprotected) edges, in which the cross-section of the metal layers was directly in contact with the corrosive atmospheres. This scheme is a common practice in durability research works [32] to properly study the influence of the weathering agents in both undamaged and pre-damaged edges.

Silvered mirrors were composed on the top by glass or polymer surfaces that protect the silver layer of possible environmental weathering. Then, the reflector layer was covered on the back side by a copper layer and different protective paints which prevent the corrosion penetration. The thickness and composition of these paint layers played an important role in order to determine the durability against the corrosion. Due to environmental reasons, current research efforts are ongoing to reduce or remove the lead content of the paints [33].

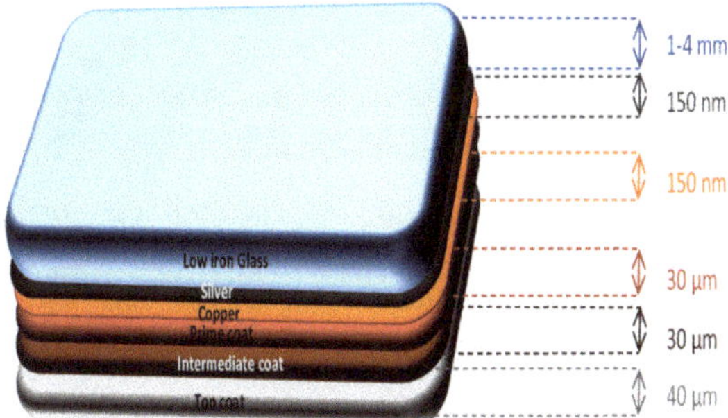

Figure 1. Schematic composition of silvered-glass.

A variety of reflector samples from a number of manufacturers and subjected to two durability tests (Copper-accelerated acetic acid salt spray (CASS) and Kesternich tests, see Section 2.3) were selected for this study, to be analyzed afterwards with the two measurement techniques. All the samples subjected to the CASS test and were labelled with a "C" followed by a number, from one to nine. For the Kesternich tests, the samples were labelled with a "K" followed by a number, from one to nine. Table 1 shows the main characteristics of all the samples tested, including the number and thickness of the back paint layers and the initial reflectance values. Regarding the samples K-1 to K-3, initial damage (a scratch on the paint) was willfully done in order to expose the entire reflector layer to the Kesternich atmosphere [10].

Table 1. Main characteristics of the samples tested.

Sample Code	Number of Protected Edges	Number of Back Paint Layers	Thickness of the Back Paint Layers (µm)	Initial $\rho_{\lambda,\varphi}$ (–)	Initial $\rho_{s,h}$ (–)
C-1	4	3	32-40-30	0.964	0.951
C-2	4	3	32-40-30	0.964	0.951
C-3	4	3	32-40-30	0.964	0.953
C-4	1	2	30-30	0.967	0.955
C-5	1	3	28-35-35	0.961	0.951
C-6	1	3	28-35-35	0.964	0.951
C-7	1	2	30-30	0.966	0.955
C-8	1	3	28-35-35	0.961	0.951
C-9	1	2	30-30	0.965	0.954
K-1	1	3	28-37-37	0.954	0.945
K-2	1	3	28-37-37	0.956	0.944
K-3	1	3	28-37-37	0.956	0.944
K-4	4	3	28-37-37	0.957	0.944
K-5	4	3	28-37-37	0.957	0.943
K-6	4	3	28-37-37	0.958	0.945
K-7	4	3	28-37-37	0.959	0.944
K-8	4	3	28-37-37	0.959	0.945
K-9	4	3	28-37-37	0.958	0.943

2.3. Durability Tests

All the samples included in this study have in common that the main degradation mechanism provoked by the two durability experiments applied (CASS and Kesternich tests) was the corrosion in the reflective silver layer, both in the form of corrosion spots or corrosion penetration through the edges. This is one of the typical degradation effect reported for silvered-glass reflectors [13,34].

On the one hand, CASS test is one of the most common aging experiments applied to simulate corrosion in solar reflectors [13]. In accordance with the ISO 9227 [16], the samples were tested at a temperature (T) of $T = 50$ °C and 100% of relative humidity (RH), where an uninterrupted spray composed by a solution of demineralized water, 50 g/L of NaCl and 0.26 g/L of $CuCl_2$ was continuously wetting the samples. The condensation rate obtained for a surface of 80 cm^2 in the testing chamber with these conditions was 1.5 mL/h. Moreover, the pH of the solution was kept between 3.1 and 3.3 and it was adjusted by adding acid or basic compounds such as HCl or CH_3COOH and NaOH. Three CASS were conducted, varying the testing time (which was 330, 430, and 480 h), with the goal of producing several different levels of corrosion.

On the other hand, Kesternich test was utilized to reproduce industrial atmospheres where the concentration of corrosive gases is very high (typically known as acid rain conditions) and consequently the main degradation mechanism is the corrosion of the reflective layer. To simulate such polluted environments, the samples were subjected to two different SO_2 concentrations (3333 and 6667 ppm) and temperatures ($T = \{25, 40, 50\}$ °C) during 8 h and to ambient conditions for 16 h, which suppose a cycle of 24 h in total. As in the previous test, different testing time was applied to vary the severity of the corrosion.

In addition to the testing conditions, the intensity of the corrosion appearing in the different samples tested depended on the number, the thickness, and the composition of the back coating layers (see Table 1). Table 2 presents a summary of the testing conditions corresponding to the different experiments applied.

Table 2. Tests conditions of the experiments applied to the studied reflector samples.

Sample Code	Durability Test	Testing Conditions	Testing Time (h)
C-1, C-2, C-3		$T = 50 \pm 2$ °C, pH = [3.1, 3.3] at 25 °C	480
C-4, C-5, C-6, C-7	CASS	Sprayed NaCl solution of 50 ± 5 g/L and	430
C-8, C-9		0.26 ± 0.02 g/L $CuCl_2$	330
K-1, K-2, K-3		$T = 40$ °C, $RH = 100\%$ $[SO_2]$ gas = 6667 ppm	910
K-4, K-5, K-6	Kesternich	$T = 50$ °C, $RH = 100\%$ $[SO_2]$ gas= 3333 ppm	768
K-7, K-8, K-9		$T = 25$ °C, $RH = 100\%$ $[SO_2]$ gas = 3333 ppm	720

2.4. Analysis Techniques

This section describes the two optical measurement equipment used to measure reflectance in both the conventional and the improved techniques.

2.4.1. Reflectometer

The equipment selected to measure $\rho_{\lambda,\varphi}$, was the portable specular reflectometer 15R-USB by Devices and Services (D&S, Dallas, TX, USA) [35], which was specifically developed by the company in cooperation with Sandia National Laboratories to assess solar reflectors [36] (see Figure 2). It has a light-emitting diode (LED) source of $\lambda = [635, 685]$ nm, with a peak at $\lambda = 660$ nm. φ can be selected from $\varphi = \{3.5, 7.5, 12.5, 23.0\}$ mrad, and the $\theta_i = 15°$. The instrument produces a collimated beam to a diameter of 10 mm (which corresponds to the measurement spot size) so that all of the reflected beam can be collected by the 22 mm diameter receiver lens. The beam deviation of the collimated incident beam is $\varphi_i = 5$ mrad and therefore almost matching the sun disc on a clear-sky day. The instrument allows measuring curved mirrors and also first and second surface mirrors with different top-layer thickness.

Figure 2. Portable specular reflectometer 15R-USB by Devices and Services (D&S).

All the measurements were taken at $\varphi = 12.5$ mrad. The instrument used (serial number 117) has a repeatability of ±0.002 and a resolution of ±0.001. The calibrated reference standard, a 4-mm second-surface silvered-glass sample by OMT (serial number OMT-214044-02), has an uncertainty of 0.0015. Considering these values, the expanded type B uncertainty is 0.006. All measurements were taken in steady conditions (at constant temperature) by the same technician.

2.4.2. Spectrophotometer

The scanning spectrophotometer (model Lambda 1050, Perkin Elmer (PE), Waltham, MA, USA) with a 150-mm diameter integrating-sphere accessory was employed to measure $\rho_{\lambda,h}$ (see Figure 3) [37]. It is a UV/Vis/NIR double beam and double monochromator instrument with two light sources, a deuterium lamp for the UV range, and a halogen lamp for the Vis/NIR range. The beam source has a spectral λ range from 175 to 3300 nm, $\theta_i = 8°$ and is about 17 mm by 9 mm size. The detection system is composed of a photomultiplier for the UV/Vis range, and two detectors (a Peltier controlled PbS detector and a InGaAs detector) for the NIR range. The measurement spot in this case is 9 mm by 17 mm.

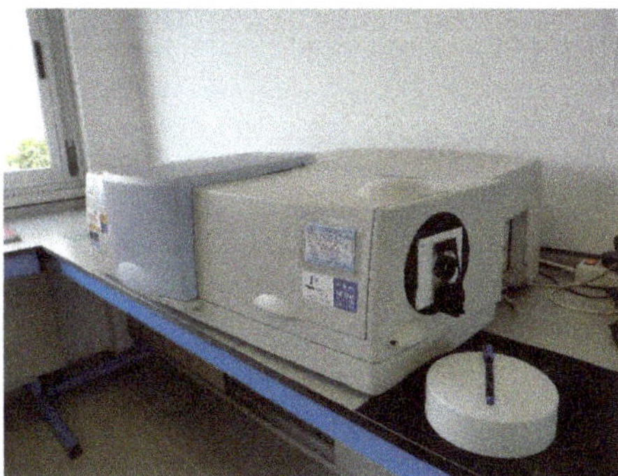

Figure 3. Spectrophotometer Lambda 1050 by Perkin Elmer (PE).

Measurements were taken in 5 nm steps in the range $\lambda = [320, 2500]$ nm. The maximum response time (0.04 s) was selected in the whole λ range, except in $\lambda = [600, 880]$ nm, where the minimum response time (1 s) was chosen. This combined response time was selected because it presents a proper compromise between accuracy and measuring time. The instrument used (serial number 1050N9061802) has an accuracy of ±0.007 (at 635 nm). The calibrated reference standard, a 4-mm second-surface silvered-glass sample by OMT (serial number OMT-214044-02), has an uncertainty of

0.0015. Considering these values, the expanded type B uncertainty is 0.016. All measurements were taken in steady conditions (at constant temperature and with an opaque cover) by the same technician.

2.4.3. EDX

Energy dispersive X-ray analysis (EDX) is a technique that uses X-ray to identify the elemental composition of a material. In our case, a QUANTAX EDS system (Bruker, Durham, UK) was utilized to perform the analysis. This EDX system is coupled to a scanning electron microscopy (SEM), model S-3400N (Hitachi, Tokyo, Japan) that generates a microscopic image that is scanned by the EDX system. The results provided by the EDX analysis consist of spectra which exhibit peaks corresponding to the elemental composition of the sample analyzed.

2.5. Conventional Reflectance Measurement Technique

The conventional method applied is in agreement with the current version of the "SolarPACES Reflectance Guideline" [15], which has been extensively employed to characterized solar reflectors in durability studies [37–39]. According to it, the following parameters must be measured to monitor the aging of solar reflectors:

- $\rho_{s,h}$ at $\theta_i \leq 15°$ and in the range $\lambda = [320, 2500]$ nm.
- $\rho_{\lambda,\varphi}$ at one defined λ in the range $\lambda = [400, 700]$ nm, $\theta_i \leq 15°$, and a φ in the range of $\varphi = [0, 20]$ mrad.
- $\rho_{\lambda,h}$ at the same λ as for $\rho_{\lambda,\varphi}$ to calculate the specularity of the reflector samples.

As it is recommended in this guideline, $\rho_{\lambda,\varphi}$ was measured in five points of each reflector sample, taken in the center of the sample and close to the four corners of the sample (see Figure 4), and the average value as well as the standard deviation was reported. The measurements were always taken at a distance to the sample edge higher than 10 mm. The equipment used for the measurements was the portable specular reflectometer 15R-USB by D&S, described in Section 2.4.1. A mask was always used to characterize each reflector sample in order to always measure the same spots and so properly assess the evolution of the possible corrosion partners.

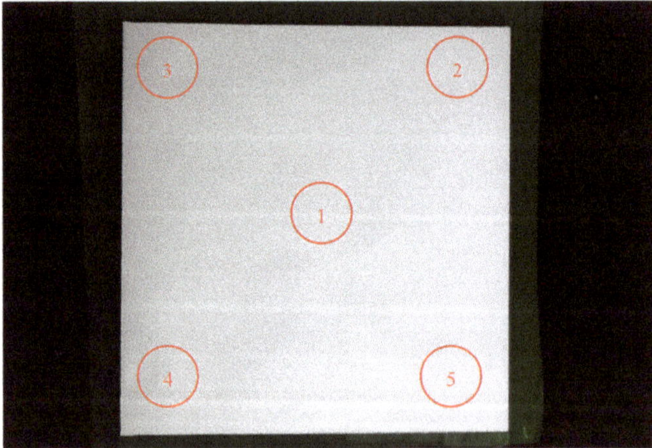

Figure 4. Five spots measured on each reflector sample with the reflectometer, according to the conventional measurement technique.

In addition, $\rho_{s,h}$ measurements were taken at the center of the reflector samples and repeated at the same point thrice (rotating the sample 0°—no rotation—, 90°, and 180°) to check any possible anisotropy. Due to the shape of the beam spot and the possible inaccuracy in the positing after the rotation, the three measurement spots might not be exactly the same (see Figure 5). Again, both the average value and the standard deviation were reported. The equipment used for the measurements was the spectrophotometer Lambda 1050 by PE, described in Section 2.4.2.

Figure 5. Three spots measured on each reflector sample with the spectrophotometer, according to the conventional measurement technique.

As it has been stated, with this method the spots monitored are always the same and consequently it is impossible to distinguish between measurements taken in corroded and non-corroded areas. This fact can cause misunderstanding and errors in the average reflectance of the whole reflector sample because in a reflector whose corroded area is almost negligible, this small degraded area might coincide with the measurement spot and consequently the reflectance would decrease drastically in an unfair manner. Also, the opposite event could happen, in a reflector sample whose useful surface is quite small because the corrosion affects most of the surface, the measurement spot could coincide with a non-corroded area and the reflectance would be much higher than the real one. For this reason, in many occasions the results achieved with this method are not representative of the average reflectance of the whole reflector sample.

2.6. Improved Reflectance Measurement Technique

An improved reflectance measurement technique was developed in this study to obtain more representative results by taking into consideration the portion of the total reflector area, A_T, affected by the corrosion. Following this goal, this method is based on distinguishing the corroded area, A_C, and the useful (non-corroded) area, A_{NC}, of the reflector sample, and weighting the average reflectance of each area accordingly, as indicated in Equation (2) for the weighted monochromatic specular reflectance, $\rho_{\lambda,\varphi,w}$, and Equation (3) for the weighted solar hemispherical reflectance, $\rho_{s,h,w}$.

$$\rho_{\lambda,\varphi,w} = \frac{A_{NC}}{A_T} \times \rho_{\lambda,\varphi,NC} + \frac{A_C}{A_T} \times \rho_{\lambda,\varphi,C} \tag{2}$$

$$\rho_{s,h,w} = \frac{A_{NC}}{A_T} \times \rho_{s,h,NC} + \frac{A_C}{A_T} \times \rho_{s,h,C} \tag{3}$$

where the subscripts "NC" and "C" in the reflectance mean non-corroded and corroded, respectively. To calculate the weighted reflectance, the first step consists in taking an image of the sample to determine A_C and A_{NC}.

In Figure 6, it is depicted the histogram with the different tonalities of black and white that could exist in the pixels (from zero to 255). Depending on the threshold chosen, the detection of black and white can vary. If the threshold selected is zero, all the pixels colors will be detected as white, while if it is 255, the opposite case would occur, and all the pixels will be detected as black. The Figure 6a shows the histogram of a sample with black corrosion. In this case, the white tonalities started in the range of 188 to 232 and the black from 58 to 125. The threshold selected was 148, distinguishing perfectly between corroded (from zero to 148) and non-corroded area (from 148 to 255). However, for samples where the corrosion appears with a color different to black, the threshold should be modified in order to detect this corrosion (see Figure 6b). In this case, the corrosion tonalities are yellow and the threshold selected to recognize this color as corrosion was 194. As a consequence, the corroded area was counted from zero to 194.

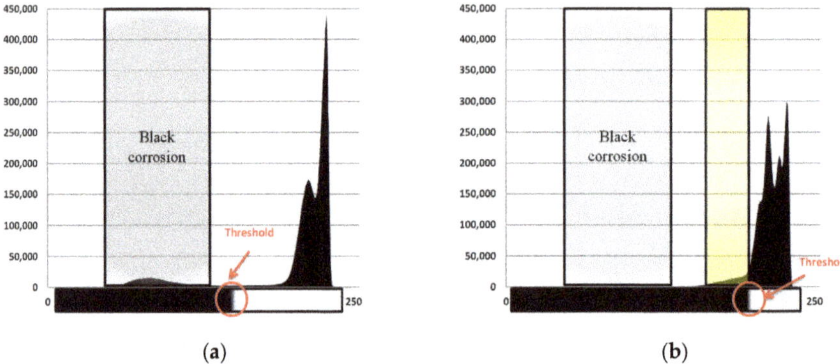

(a) (b)

Figure 6. Histogram image of a corroded sample with black corrosion (**a**) and yellow corrosion (**b**).

The pictures were taken with a homogenous illumination of 1000 mA in the whole sample to avoid shadow areas and with an exposure time of 1/15 s. As it was previously explained, for typical corrosion whose corrosion color is black, the threshold employed was 148. Nevertheless, for corrosion whose color is not black, for instance yellow, the threshold should be changed to higher values in order to detect the degradation. Additionally, a distinction between the corroded area starting in the cut edges and originated by corrosion spots can be implemented. With an image treatment software (Matlab or ImageJ), it is possible to perform the transformation of the real image to a binary image, where the corroded surface is in black and the useful area is in white, permitting to estimate the no-corroded surface percentage (see Figure 7). After several tests, it was determined that the minimum quality recommended to take the image should be 254.6 pixels/mm, which corresponds to a vertical and horizontal resolution of 300 ppi. The accuracy of the software is good enough because the operator of the software can avoid manually the side effects that could be interpreted as corrosion but they are not. Examples of this kind of side effects are shading when the photo was taken, and soiling deposited on the reflector surface. Consequently, the accuracy in detecting the real corrosion is very high. To have a quantitative value of this accuracy, one corroded reflector sample was analyzed by taken five pictures and calculating the corroded area with the software, keeping all the parameters constant. The repeatability obtained was ±0.07%.

Figure 7. Binary image of Figure 8 to calculate the total corroded area.

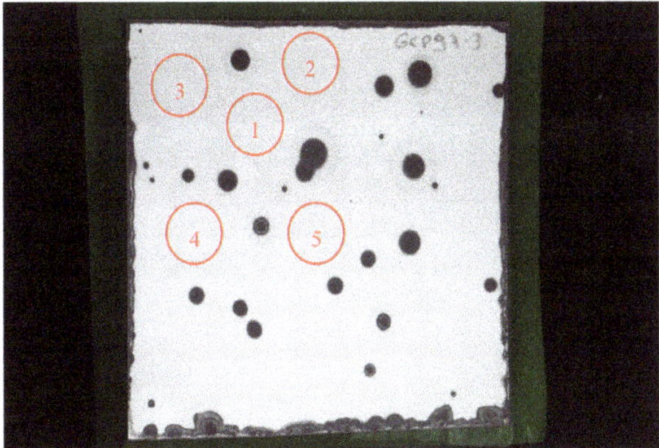

Figure 8. Five measurements taken along the non-corroded surface of a partially corroded reflector.

The second step is based on taking the reflectance measurements in the useful area to calculate $\rho_{\lambda,\varphi,NC}$ or $\rho_{s,h,NC}$. On the one hand, for corrosion with a dark color, the reflectance of the corroded area can be approximated to zero (black surface), $\rho_{\lambda,\varphi,C} = \rho_{s,h,C} = 0$. Then, for the calculation of $\rho_{\lambda,\varphi,NC}$ with the reflectometer, five measurements are randomly taken along the non-corroded reflector surface. In samples where the corrosion is heterogeneous, special care must be taken to avoid measuring in corrosion spots and in the corroded edges (see Figures 8–10). On the other hand, if the corrosion is not dark (for instance, yellow), it is necessary to measure, as a minimum, once in the corroded area to calculate the $\rho_{\lambda,\varphi,C}$ or $\rho_{s,h,C}$ and to change the threshold value for taking into consideration this area (see Figure 10). The same rules should be followed to calculate $\rho_{s,h,NC}$ with the spectrometer, with the only exception that in this case only three measurements are recommended to avoid a very time consuming process.

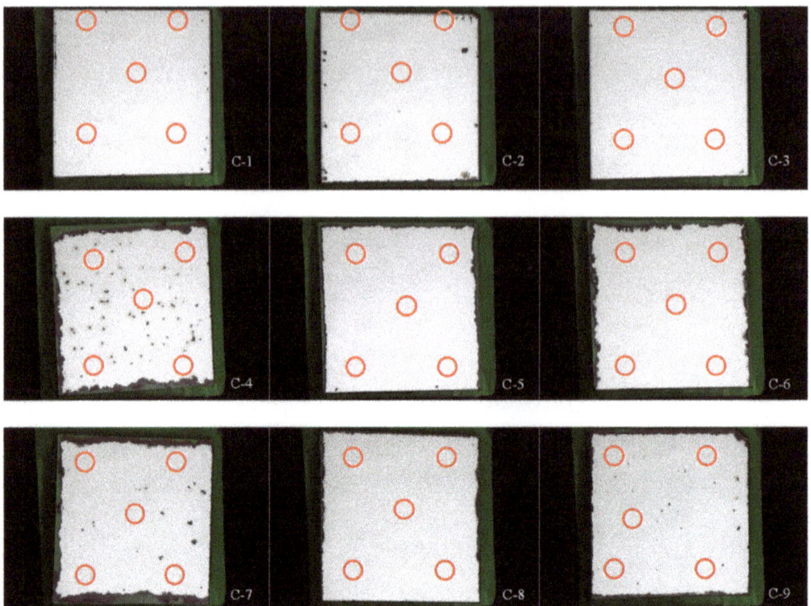

Figure 9. Images of the corroded reflector samples aged in the CASS test, with the measurement spots marked with red circles.

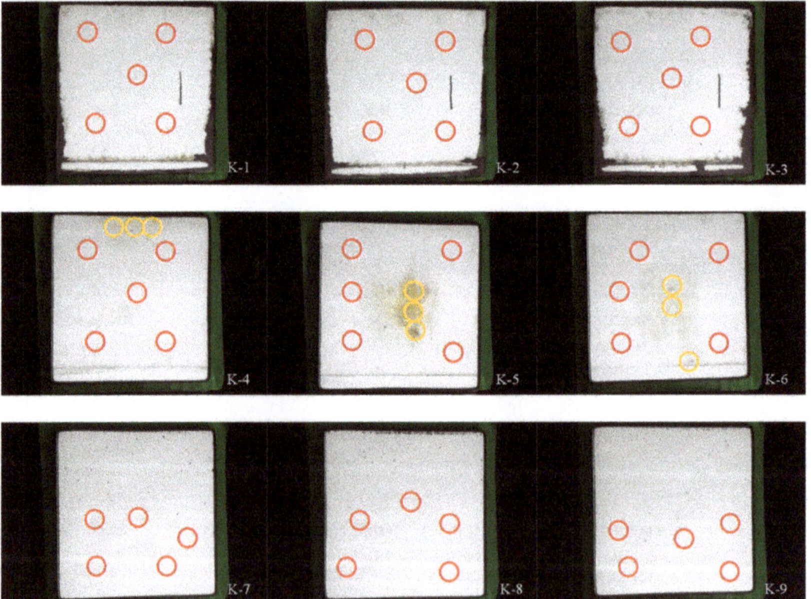

Figure 10. Images of the corroded reflector samples tested in the Kesternich test, with the measurement spots marked with red circles (non-corroded area) or yellow circles (corroded area).

3. Results and Discussion

This section sets out results of the corrosion products formed, the corrosion rate and the reflectance (both average and standard deviation values) of corroded samples measured with the two different measurement techniques. For purposes of clarity, the results are divided into two subsections according to the accelerated aging test used to corrode the reflectors. Moreover, images of these reflector samples are provided to show the different corrosion levels and the points where $\rho_{\lambda,\varphi,C}$ (in yellow) and $\rho_{\lambda,\varphi,NC}$ (in red) were measured.

3.1. Reflectance Analysis

3.1.1. CASS Test

Table 3 presents the average and standard deviation of $\rho_{\lambda,\varphi}$, $\rho_{\lambda,\varphi,NC}$, $\rho_{s,h}$, and $\rho_{s,h,NC}$ of the corroded samples using both measurement techniques and after several testing times in the CASS test (see Table 2 for the testing conditions). The values of $\rho_{\lambda,\varphi,w}$ and $\rho_{s,h,w}$ are calculated through the Equations (1) and (2), weighting the $\rho_{\lambda,\varphi,NC}$ and $\rho_{s,h,NC}$ with the corroded area, as also shown in the table. The weighted reflectance value obtained with the improved measurement technique and the reflectance measured with the conventional method are both in bold for an easy comparison of the results (see the Discussion section). Furthermore, the corresponding images of these samples are illustrated in Figure 9.

Table 3. Results of the corroded area and the monochromatic specular and solar hemispherical reflectance values for both reflectance measurement techniques, applied to the corroded samples aged in the CASS test.

Sample Code	Corroded Area (%)	Monochromatic Specular Reflectance						Solar Hemispherical Reflectance					
		Conventional Reflectance Method		Improved Reflectance Method				Conventional Reflectance Method		Improved Reflectance Method			
		$\rho_{\lambda,\varphi}$ (–)		$\rho_{\lambda,\varphi,NC}$ (–)		$\rho_{\lambda,\varphi,w}$ (–)		$\rho_{s,h}$ (–)		$\rho_{s,h,NC}$ (–)		$\rho_{s,h,w}$ (–)	
		$\overline{X}$	σ	$\overline{X}$	σ	$\overline{X}$	σ	$\overline{X}$	σ	$\overline{X}$	σ	$\overline{X}$	σ
C-1	0.30	**0.964**	0.001	0.963	0.001	**0.960**	0.001	**0.950**	0.000	0.951	0.001	**0.948**	0.001
C-2	1.42	**0.963**	0.001	0.962	0.000	**0.949**	0.000	**0.949**	0.004	0.951	0.000	**0.937**	0.000
C-3	0.11	**0.964**	0.001	0.964	0.001	**0.963**	0.001	**0.952**	0.000	0.952	0.000	**0.951**	0.000
C-4	13.00	**0.921**	0.040	0.945	0.012	**0.822**	0.010	**0.931**	0.008	0.935	0.007	**0.813**	0.006
C-5	7.72	**0.960**	0.002	0.959	0.000	**0.886**	0.000	**0.951**	0.000	0.951	0.000	**0.878**	0.000
C-6	9.45	**0.963**	0.001	0.964	0.000	**0.873**	0.000	**0.951**	0.002	0.951	0.001	**0.861**	0.001
C-7	15.00	**0.963**	0.001	0.963	0.000	**0.819**	0.000	**0.950**	0.001	0.951	0.000	**0.808**	0.000
C-8	3.80	**0.959**	0.003	0.959	0.002	**0.922**	0.002	**0.949**	0.000	0.949	0.000	**0.913**	0.000
C-9	7.00	**0.959**	0.003	0.964	0.001	**0.892**	0.001	**0.943**	0.000	0.949	0.000	**0.883**	0.000

3.1.2. Kesternich Test

Regarding the samples subjected to the Kesternich test (see Table 2 for the testing conditions), Table 4 presents the average and standard deviation of $\rho_{\lambda,\varphi}$, $\rho_{\lambda,\varphi,NC}$, $\rho_{s,h}$, and $\rho_{s,h,NC}$ for the measurements using both measurement techniques after several times in three Kesternich tests at different gas concentration and temperature values (see Table 2 for the specific testing conditions). For the material K-4, K-5, and K-6, the $\rho_{\lambda,\varphi,C}$ obtained was 0.85, 0.32, and 0.86 ppt and the $\rho_{s,h,C}$ was 0.87, 0.82, and 0.87 ppt, respectively. $\rho_{\lambda,\varphi,C}$ and $\rho_{s,h,C}$ for the rest of the samples were zero. Also, the corroded area is presented in this table to calculate the weighted reflectance ($\rho_{\lambda,\varphi,w}$ and $\rho_{s,h,w}$) with the improved method. As in Table 3, final reflectance values of both methods are shown in bold. In addition, Figure 10 exhibits representative images of the corroded samples after the Kesternich test.

Table 4. Results of the corroded area and the monochromatic specular and solar hemispherical reflectance values for both reflectance measurement techniques, applied to the corroded samples aged in the Kerternich test.

Sample Code	Corroded Area (%)	Monochromatic Specular Reflectance						Solar Hemispherical Reflectance					
		Conventional Reflectance Method		Improved Reflectance Method				Conventional Reflectance Method		Improved Reflectance Method			
		$\rho_{\lambda,\varphi}$ (–)		$\rho_{\lambda,\varphi,NC}$ (–)		$\rho_{\lambda,\varphi,w}$ (–)		$\rho_{s,h}$ (–)		$\rho_{s,h,NC}$ (–)		$\rho_{s,h,w}$ (–)	
		$\overline{X}$	σ	$\overline{X}$	σ	$\overline{X}$	σ	$\overline{X}$	σ	$\overline{X}$	σ	$\overline{X}$	σ
K-1	9.00	0.954	0.001	0.954	0.001	0.869	0.001	0.945	0.000	0.943	0.000	0.858	0.000
K-2	9.58	0.951	0.003	0.953	0.001	0.862	0.001	0.944	0.000	0.942	0.000	0.852	0.000
K-3	12.78	0.954	0.002	0.955	0.001	0.833	0.001	0.944	0.001	0.942	0.000	0.822	0.000
K-4	4.32	0.955	0.003	0.957	0.002	0.952	0.002	0.944	0.000	0.941	0.001	0.938	0.001
K-5	8.82	0.778	0.300	0.932	0.030	0.879	0.027	0.782	0.020	0.939	0.004	0.929	0.004
K-6	7.02	0.929	0.040	0.960	0.003	0.953	0.003	0.887	0.010	0.943	0.002	0.938	0.002
K-7	0.37	0.954	0.002	0.957	0.001	0.953	0.001	0.945	0.000	0.945	0.000	0.942	0.000
K-8	0.95	0.954	0.002	0.955	0.001	0.946	0.001	0.944	0.000	0.945	0.000	0.936	0.000
K-9	0.43	0.954	0.001	0.956	0.000	0.951	0.000	0.942	0.001	0.944	0.000	0.940	0.000

3.1.3. Discussion of the Measurement Techniques

As can be seen in the Tables 3 and 4, the reflectance results of both measurement techniques are quite different. Usually, the conventional method provides much higher reflectance values than the improved method because the corroded area of the sample is not properly considered.

On the one hand, if $\rho_{\lambda,\varphi}$ and $\rho_{\lambda,\varphi,NC}$ are compared to $\rho_{s,h}$ and $\rho_{s,h,NC}$, the values are quite similar for most of the cases (with differences below the type B uncertainty of the equipment) because the probability of measuring in non-corroded area with the conventional method is high, due to the low portion of corroded area (not greater than the 15% of the total area), as can be observed in the tables and pictures. However, given that the conventional method always measures in the same spots, there is a chance that the corrosion may appear in the measurement area. For instance, this event occurs in samples C-4 and K-5 for the monochromatic specular reflectance (where the measurement spots are distributed in the whole surface, see Figure 4), and K-5 and K-6 for the solar hemispherical reflectance (where the measurements spots are located in the central area, see Figure 5). This fact also provokes an increase in the standard deviation (reaching values up to 0.300). Additionally, the $\rho_{\lambda,\varphi,NC}$ and $\rho_{s,h,NC}$ assess the quality of the mirror surface which is not corroded, providing a more reliable reflectance average because uncertainty related to the corroded area is reduced. Thus, with these parameters ($\rho_{\lambda,\varphi,NC}$ and $\rho_{s,h,NC}$) it is possible to evaluate the quality of the reflector layer.

On the other hand, a great discrepancy between $\rho_{\lambda,\varphi}$ and $\rho_{\lambda,\varphi,w}$, and $\rho_{s,h}$ and $\rho_{s,h,w}$ exists for most of the reflectors, where the improved measurement technique provides smaller reflectance results because the corroded area is taken into consideration. To quantify the importance of this discrepancy, the reflectance differences between the conventional and the improved methods, $\rho_{\lambda,\varphi}$-$\rho_{\lambda,\varphi,w}$ and $\rho_{s,h}$-$\rho_{s,h,w}$, are calculated (for all samples except K-5 and K-6). As a result, the average values achieved are 0.053 ppt for both the monochromatic specular and solar hemispherical reflectance. Especially, the samples C-4, C-6, C-7 (Table 3), and K-3 (Table 4) show significant monochromatic specular reflectance discrepancies, where the maximum difference of 0.144 ppt is reached for the sample C-7. Regarding the solar hemispherical reflectance, a maximum reflectance discrepancy of 0.142 ppt between both methods is achieved for the sample C-7. This fact occurs because these specific samples are very damaged, as it is depicted in the Figures 9 and 10. Consequently, the measurements taken with the conventional method only were taken in the non-corroded area, obtaining a reflectance value much higher than the real. Otherwise, the opposite case occurs for the samples K-5 and K-6 where the monochromatic specular and solar hemispherical reflectance in the improved method is 0.101, 0.024 and 0.147, 0.051 ppt higher than in the conventional one, respectively. This difference is induced by a coincidence between the measurement spots of the conventional method and the corroded area, provoking an excessive decrease of the reflectance.

Finally, there are two cases that are worth mentioning (C-7 and K-3), where the main corrosion is suffered in the edges of the reflector samples. In both cases, the corroded area is very high but the two types of measured reflectance ($\rho_{\lambda,\varphi}$ and $\rho_{\lambda,\varphi,NC}$ and $\rho_{s,h}$ and $\rho_{s,h,NC}$) show unrealistically high values. However, the weighted reflectance ($\rho_{\lambda,\varphi,w}$ and $\rho_{s,h,w}$) do present low values, which are more reliable because the corrosion mainly appeared in areas not accessible for the measurement instruments (the edges).

3.2. Corrosion Products Formed and Corrosion Rates

This section presents all the results of the advanced analysis performed in the corroded samples in order to determine the corrosion rates and the corrosion products appeared after the accelerated aging tests. It is divided into two sections, according to the accelerated aging tests.

3.2.1. CASS Test

As depicted in Figure 9, the corrosion of the samples in the CASS test was provoked by two different degradation mechanisms, which are the occurrence of corrosion spots and the corrosion along the edges. In order to determine the corrosion products that might appear in both types of degradation defects, an EDX analysis was performed in the silver layer of a reflector sample in the initial status and also in a corrosion defect of a reflector sample after being subjected to the CASS test during 480 h. For both types of degradation defects, the formation of oxides and chlorides of silver was observed (see Figure 11). This originates the appearance of new compounds whose reflectance is zero, subsequently reducing the optical properties of the reflectors. As can be seen, the EDX analysis before the CASS test did not detect the presence of Cl^- ions in the composition of the silver layer (see Figure 11a). However, after 480 h of the CASS test, a new peak appeared in the spectrum, which corresponds to Cl^- ion (see Figure 11b). The presence of silicon, magnesium, and sodium is due to the proximity of the glass to the measured zone. In addition, copper is detected because this layer is deposited on the back side of the silver layer for protection. Lastly, the non-identified peaks correspond to the gold which was sputtered on the sample to improve the quality of the EDX analysis.

Another important parameter to assess the corrosion detected in the samples is the corrosion rate (see Table 5), which is calculated as the corroded area divided by the testing time. This parameter varied in function of the test severity (see Table 2), the number of protected edges and the types of paints of the reflector (see Table 1).

Table 5. Corrosion rates of samples tested in the CASS test.

Sample Code	C-1	C-2	C-3	C-4	C-5	C-6	C-7	C-8	C-9
Corrosion Rate (cm²/h)	0.0006	0.003	0.0002	0.0300	0.0220	0.0200	0.0500	0.012	0.0210

The main conclusions obtained from the corrosion rates are:

- The protection of the edges is essential to improve the durability of a reflector. As it is shown in the Table 5, the samples tested during the longest time obtained the lowest corrosion rate because all of the edges were protected.
- As for the number of paint layers, the samples C-4, C-5, C-6, and C-7 were tested during the same time (430 h). However, a great difference is observed among their corrosion rates, being higher for samples C-4 and C-7, which contain two paint layers instead of three. The same argument is used for samples tested during 330 h, where C-9 (two paint layers) has a corrosion rate twice as high as C-8 (three paint layers).
- Another parameter to consider in the evaluation of the corrosion rate is the testing time. As can be seen for samples with only one protected edge and the same number of paint layers, the corrosion

rate increases with time. For example, this is the case of C-4 and C-7 against C-9 (both with two paint layers), where C-7 has higher corrosion rate due to the longer testing time.

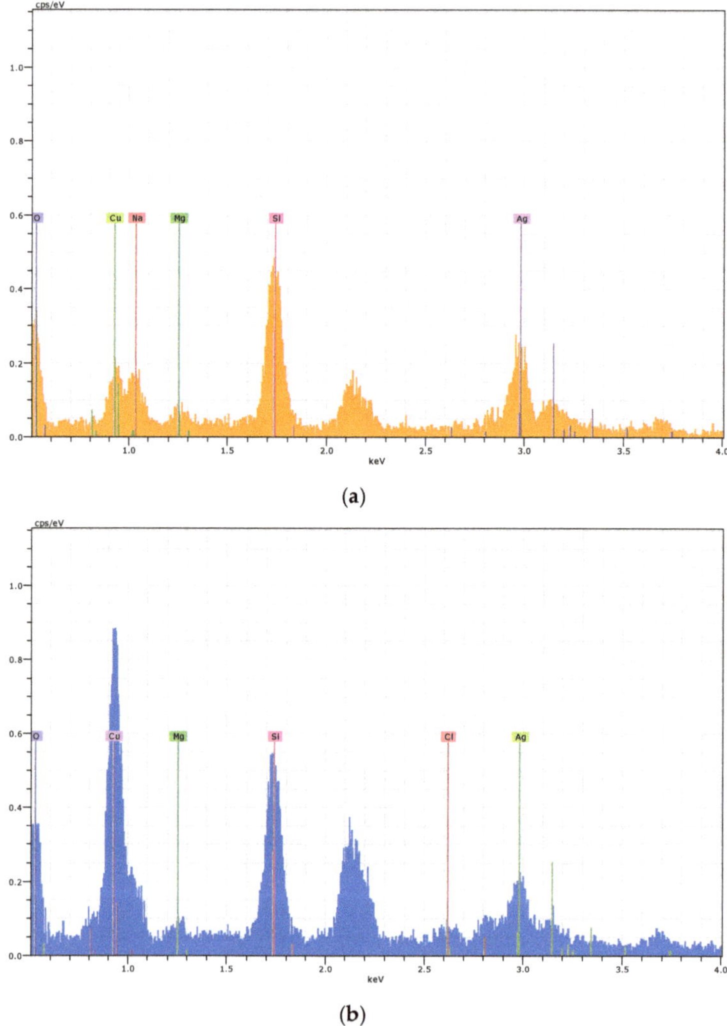

Figure 11. Energy dispersive X-ray analysis (EDX) analysis of the reflector silver layer before (**a**) and after of 480 h (**b**) of the CASS test.

It is also important to highlight that these conclusions can be also extrapolated to the differences detected in the corroded area (see Table 3) because, as expected, a perfect match was found between both parameters (corroded area and corrosion rate).

3.2.2. Kesternich Test

As was previously reported for the CASS test, the main corrosion mechanisms are the appearance of corrosion spots and edge corrosion. To characterize the products formed in the corrosion process, an EDX analysis of a sample tested during 910 h in a Kesternich test was carried out. It was observed that the silver layer was significantly affected by the SO_2 atmosphere, due to the reaction of the silver with the sulfur, and provoking the appearance of new compounds. As illustrated in Figure 12, the emergence of sulfur is detected, as well as other compounds such as silicon, gold, and sodium (aforementioned and explained in Section 3.2.1). Another remarkable result is the absence of copper in this analysis, probably due to a total corrosion of the copper layer by the sulfurous atmosphere.

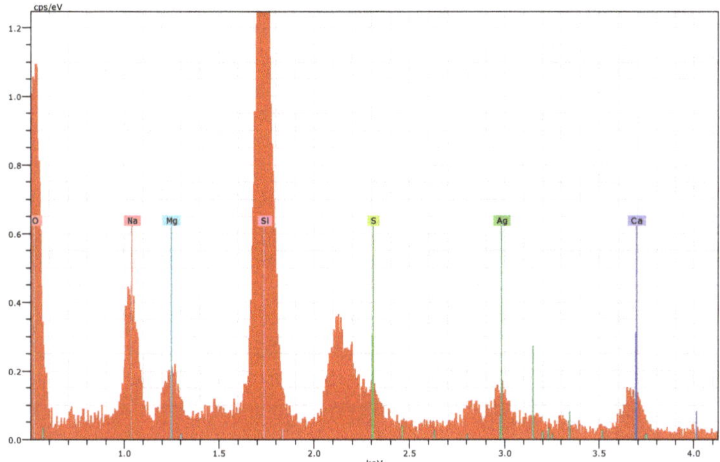

Figure 12. EDX analysis of the reflector silver layer after of 910 h of Kesternich test.

In addition, the corrosion rates of all the samples tested in the Kesternich test were obtained (see Table 6). The method to calculate this parameter was already described in Section 3.1.

Table 6. Corrosion rates of samples tested in the Kesternich test.

Sample Code	K-1	K-2	K-3	K-4	K-5	K-6	K-7	K-8	K-9
Corrosion Rate (cm^2/h)	0.010	0.010	0.014	0.005	0.010	0.009	0.001	0.001	0.001

In this test, the influence of the temperature and the SO_2 concentration must be taken into account, arriving at the following main considerations:

- As it was mentioned in the CASS results, the protection of the edges is a very important issue. This was also noticed in the Kesternich test because the three samples with only one protected edge (K-1, K-2, and K-3) showed the highest corrosion rates.
- With respect to the gas concentration, it was also noticed that the degradation is increased at the highest SO_2 concentration (samples K-1, K-2, and K-3).
- Finally, the influence of the testing temperature can be analyzed by comparing samples tested at the same SO_2 concentration (3333 ppm) but at different temperatures, i.e., samples K-4, K-5, and K-6 (tested at 50 °C) against samples K-7, K-8, and K-9 (tested at 25 °C). As can be observed, the corrosion rates are much higher for samples tested at the highest temperature. Consequently, the influence of the temperature is crucial in this test.

As it was already mentioned for the CASS test, these conclusions can be also extrapolated to the differences detected in the corroded area (see Table 4) because, as expected, a perfect match was found between both parameters (corroded area and corrosion rate).

4. Conclusions

The investigation presented in this work demonstrates that the reflectance of corroded reflector samples is not suitably assessed by the conventional measurement technique. Thus, a new methodology is required to analyze corroded reflectors. Following up on this goal, an improved measurement technique that provides an appropriate reflectance value for corroded samples was developed. This method was able to distinguish between corroded and non-corroded surfaces and to determine a weighted reflectance value. The average differences between both methods are 0.053 ppt for both the monochromatic specular and the solar hemispherical reflectance. Discrepancies of up to 0.144 and 0.147 ppt are reached for monochromatic specular and solar hemispherical reflectance, respectively. As it was demonstrated by image inspection of the reflector samples, the reliability of both monochromatic specular and solar hemispherical reflectance was much higher when the improved method was utilized, even in the cases where the corrosion is mainly found near the sample edges.

Moreover, it was proved that the severity of the corrosion originated in the reflectors depends on several parameters, such as the reflector material quality and the corrosion test conditions. With respect to the reflector material quality, the numbers of back paint layers as well as the number of protected edges are two of the key indicators of reflector resistance against corrosion. In this sense, it was determined that the corrosion rate and the corroded area decrease as the number of both the back paint layers and the protected edges increase. Additionally, the degradation of the reflector samples increases with the severity of the test conditions and the duration of the test.

Regarding the products formed after the aging tests, the EDX analysis of the silver layer shows that chlorine appeared in the CASS test, whereas traces of sulfur were observed after the Kesternich test. Therefore, it is concluded that the new compounds originated depend on the chemical elements added in the aging tests.

Author Contributions: Conceptualization, F.B.-M., F.S. and A.F.-G.; methodology, F.B.-M.; validation, F.B.-M.; formal analysis, F.B.-M.; investigation, F.B.-M.; resources, A.F.-G., F.S. and L.V.; data curation, F.B.-M. and A.G.-S.; writing—original draft preparation, F.B.-M.; writing—review and editing, A.F.-G., A.G.-S., F.S. and L.V.; supervision, A.F.-G., A.G.-S., F.S. and L.V.; project administration, A.F.-G., F.S. and L.V.; funding acquisition, A.F.-G., F.S. and L.V.

Funding: This work is part of the project RAISELIFE that has received funding from the European Union's Horizon 2020 research and innovation programme under grant agreement No. 686008.

Acknowledgments: Authors would like to thanks Lucía Martínez Arcos from CIEMAT as well as Tomás Jesús Reche Navarro and Carmen María Amador Cortés for their valuable contributions to carry out the tests.

Conflicts of Interest: The authors declare no conflict of interest.

References

1. Tian, C.; Feng, G.; Li, S.; Xu, F. Scenario analysis on energy consumption and CO_2 emissions reduction potential in building heating sector at community level. *Sustainability* **2019**, *11*, 5392. [CrossRef]
2. Jäger-Waldau, A. Photovoltaics and renewable energies in Europe. *Renew. Sustain. Energy Rev.* **2007**, *11*, 1414–1437. [CrossRef]
3. Montoya, F.G.; Aguilera, M.J.; Manzano-Agugliaro, F. Renewable energy production in Spain: A review. *Renew. Sustain. Energy Rev.* **2014**, *33*, 509–531. [CrossRef]
4. Eldeen, H.; Aleem, A.; Andina, D.; Baldi, S.; Barone, M.; Bruno, M.; Daniel Carletti, D.; Carrillo, R.; Carvalho, M.; Souza, S. *Advances in Renewab Energies and Power Technologies*; Elsevier: Amsterdam, The Netherlands, 2018.
5. Mohtasham, J. Review article-renewable energies. *Energy Procedia* **2015**, *74*, 1289–1297. [CrossRef]
6. Solangi, K.H.; Islam, M.R.; Saidur, R.; Rahim, N.A.; Fayaz, H. A review on global solar energy policy. *Renew. Sustain. Energy Rev.* **2011**, *15*, 2149–2163. [CrossRef]
7. Cooke, P. Green governance and green clusters: Regional & national policies for the climate change challenge of Central & Eastern Europe. *J. Open Innov. Technol. Mark. Complex.* **2015**, *1*, 1–17.
8. Ummadisingu, A.; Soni, M.S. Concentrating solar power—Technology, potential and policy in India. *Renew. Sustain. Energy Rev.* **2011**, *15*, 5169–5175. [CrossRef]
9. International Energy Agency. *Renewables Information: Overview*; IEA Stat: Paris, France, 2017; p. 3.

10. García-Segura, A.; Fernández-García, A.; Buendía-Martínez, F.; Ariza, M.J.; Sutter, F.; Valenzuela, L. Durability studies of solar reflectors used in concentrating solar thermal technologies under corrosive sulfurous atmospheres. *Sustainability* **2018**, *10*, 3008. [CrossRef]

11. International Energy Agency. *Technology Roadmap—Solar Thermal Electricity*; IEA Stat: Paris, France, 2014.

12. Mills, D. Advances in solar thermal electricity technology. *Sol. Energy* **2004**, *76*, 19–31. [CrossRef]

13. García-Segura, A.; Fernández-García, A.; Ariza, M.J.; Valenzuela, L.; Sutter, F. Durability studies of solar reflectors: A review. *Renew. Sustain. Energy Rev.* **2016**, *62*, 453–467. [CrossRef]

14. Fernández-García, A.; Sutter, F.; Fernández-Reche, J.; Lüpfert, E. *The Performance of Concentrating Solar Power (CSP) Systems—Modelling Measurement and Assessment*; Elsevier: Amsterdam, The Netherlands, 2017; ISBN 978-0-08-100447-0.

15. Fernández-García, A.; Sutter, F.; Montecchi, M.; Sallaberry, F.; Heimsath, A.; Heras, C.; Le Baron, E.; Soum-Glaude, A. *Parameters and Method to Evaluate Reflectance Properties of Reflector Materials for Concentrating Solar Power Technology*; Oficial Reflectance Guideline Version 3.0; SolarPACES: Almería, Spain, 2018.

16. UNE 206016. *Reflector Panels for Concentrating Solar Technologies*; UNE: Madrid, Spain, 2018.

17. Montecchi, M.; Delord, C.; Raccurt, O.; Disdier, A.; Sallaberry, F.; García de Jalón, A.; Fernández-García, A.; Meyen, S.; Happich, C.; Heimsath, A.; et al. Hemispherical reflectance results of the SolarPACES reflectance round robin. *Energy Procedia* **2015**, *69*, 1904–1907. [CrossRef]

18. Fernández-García, A.; Sutter, F.; Heimsath, A.; Montecchi, M.; Sallaberry, F.; Peña-Lapuente, A.; Delord, C.; Martínez-Arcos, L.; Reche-Navarro, T.J.; Schmid, T.; et al. Simplified analysis of solar-weighted specular reflectance for mirrors with high specularity. *AIP Conf. Proc.* **2016**, *1734*, 130006.

19. Calvo, R.; Cantero, D. A new high sensitivity and low cost solution for the measurement of reflectance loss due to dust deposition in solar collectors. In Proceedings of the SolarPACES 2013, International Conference on Concentrating Solar Power and Chemical Energy Systems, Las Vegas, NV, USA, 17–20 September 2013.

20. Wolfertstetter, F.; Pottler, K.; Geuder, N.; Affolter, R.; Merrouni, A.A.; Mezrhab, A.; Pitz-Paal, R. Monitoring of mirror and sensor soling with TraCS for improved quality of ground based irradiance measurements. *Energy Procedia* **2014**, *49*, 2422–2432. [CrossRef]

21. Bouaddi, S.; Ihlal, A.; Fernández-García, A. Soiled CSP solar reflectors modelling using dynamic linear models. *Sol. Energy* **2015**, *122*, 847–863. [CrossRef]

22. Heimsath, A.; Nitz, P. The effect of soiling on the reflectance of solar reflector materials—Model for prediction of incidence angle dependent reflectance and attenuation due to dust deposition. *Sol. Energy Mater. Sol. Cells* **2019**, *195*, 258–268. [CrossRef]

23. Sutter, F.; Meyen, S.; Heller, P.; Pitz-Paal, R. Development of a spatially resolved reflectometer to monitor corrosion of solar reflectors. *Opt. Mater.* **2013**, *35*, 1600–1609. [CrossRef]

24. Sutter, F.; Meyen, S.; Fernández-García, A.; Heller, P. Spectral characterization of specular reflectance of solar mirrors. *Sol. Energy Mater. Sol. Cells* **2016**, *145*, 248–254. [CrossRef]

25. Montecchi, M. Upgrading of ENEA solar mirror qualification set-up. *Energy Procedia* **2014**, *49*, 2154–2161. [CrossRef]

26. Heimsath, A.; Schmid, T.; Nitz, P. Angle resolved specular reflectance measured with VLABS. *Energy Procedia* **2015**, *69*, 1895–1903. [CrossRef]

27. Sutter, F.; Fernández-García, A.; Heimsath, A.; Montecchi, M.; Pelayo, C. Advanced measurement techniques to characterize the near-specular reflectance of solar mirrors. *AIP Conf. Proc.* **2019**, *2126*, 110003.

28. ISO 9050. *Glass in Building, Determination of Light Transmittance, Solar Direct Transmittance, Total Solar Energy Transmittance, Ultraviolet Transmittance and Related Glazing Factors*; International Organization for Standardization: Geneva, Switzerland, 2003.

29. ASTM G173-03. *Standard Tables for Reference Solar Spectral Irradiances: Direct Normal and Hemispherical on 37° Tilted Surface*; ASTM: West Conshohocken, PA, USA, 2003.

30. Sutter, F.; Fernández-García, A.; Heller, P.; Anderson, K.; Wilson, G.; Schumücker, M.; Marvig, P. Durability testing of silvered-glass mirrors. *Energy Procedia* **2015**, *69*, 1568–1577. [CrossRef]

31. García-Segura, A.; Fernández-García, A.; Ariza, M.J.; Sutter, F.; Diamantino, T.; Martínez-Arcos, L.; Reche-Navarro, T.J.; Valenzuela, L. Influence of gaseous pollutants and their synergistic effects on the aging of reflector materials for concentrating solar thermal technologies. *Sol. Energy Mater. Sol. Cells* **2019**, *200*, 109955. [CrossRef]

32. García-Segura, A.; Fernández-García, A.; Ariza, M.J.; Sutter, F.; Valenzuela, L. Effects of reduced sulphur atmospheres on reflector materials for concentrating solar thermal applications. *Corros. Sci.* **2018**, *133*, 78–93. [CrossRef]

33. Kennedy, C.E.; Terwilliger, K. Optical durability of candidate solar reflectors. *J. Sol. Energy Eng.* **2005**, *127*, 262–269. [CrossRef]

34. Sutter, F.; Fernández-García, A.; Wette, J.; Heller, P. Comparison and evaluation of accelerated aging tests for reflectors. *Energy Procedia* **2014**, *49*, 1718–1727. [CrossRef]

35. Sansom, C.; Fernández-García, A.; King, P.; Sutter, F.; Garcia Segura, A. Reflectometer comparison for assessment of back-silvered glass solar mirrors. *Sol. Energy* **2017**, *155*, 496–505. [CrossRef]

36. Pettit, R.B. *Characterizing Solar Mirror Materials Using Portable Reflectometers*; Sandia National Laboratories: Albuquerque, NM, USA, 1982.

37. Fernández-García, A.; Sutter, F.; Martínez-Arcos, L.; Sansom, C.; Wolfertstetter, F.; Delord, C. Equipment and methods for measuring reflectance of concentrating solar reflector materials. *Sol. Energy Mater. Sol. Cells* **2017**, *167*, 28–52. [CrossRef]

38. Wiesinger, F.; Sutter, F.; Fernández-García, A.; Reinhold, J.; Pitz-Paal, R. Sand erosion on solar reflectors: Accelerated simulation and comparison with field data. *Sol. Energy Mater. Sol. Cells* **2016**, *145*, 303–3013. [CrossRef]

39. Bouaddi, S.; Ihlal, A.; Fernández-García, A. Comparative analysis of soiling of CSP mirror materials in arid zones. *Renew. Energy* **2017**, *101*, 437–449. [CrossRef]

Article

Water Saving in CSP Plants by a Novel Hydrophilic Anti-Soiling Coating for Solar Reflectors

Johannes Wette [1,*], Aránzazu Fernández-García [2], Florian Sutter [1], Francisco Buendía-Martínez [2], David Argüelles-Arízcun [2], Itziar Azpitarte [3] and Gema Pérez [4]

[1] DLR German Aerospace Center, Institute of Solar Research, Plataforma Solar de Almería, Ctra. Senés Km. 4, 04200 Tabernas, Spain; florian.sutter@dlr.de

[2] CIEMAT-PSA, Ctra. Senés, 04200 Tabernas, Spain; arantxa.fernandez@psa.es (A.F.-G.); francisco.buendia@psa.es (F.B.-M.); david.arguelles@psa.es (D.A.-A.)

[3] IK4-Tekniker, Polo Tecnológico de Eibar, C/Iñaki Goenaga, 5. Eibar, 20600 Gipuzkoa, Spain; itziar.azpitarte@tekniker.es

[4] Rioglass Solar S.A. PI Villallana, Pola de Lena, 33695 Asturias, Spain; g.perez_ext@rioglass.com

* Correspondence: johannes.wette@dlr.de; Tel.: +34-950-362946

Received: 1 October 2019; Accepted: 4 November 2019; Published: 7 November 2019

Abstract: In this work, results of the outdoor exposure campaign of a newly developed hydrophilic anti-soiling coating for concentrated solar thermal power (CSP) mirrors are presented. The material was exposed for nearly two years under realistic outdoor conditions and the influence of two different cleaning techniques was evaluated. Mirror samples were analyzed during exposure and their reflectance and cleanliness were measured. The performance of the anti-soiling coated mirror samples was compared to conventional uncoated silvered-glass mirrors. The coatings showed appropriate anti-soiling and easy-to-clean behavior, with a mean cleanliness gain of 1 pp and maximum values under strong soiling conditions of up to over 7 pp. Cleanliness of the coated samples stayed higher throughout the whole campaign before and after cleaning, resulting in lower soiling rate compared to the reference material. Taking into account these values and supposing a threshold for cleaning of 96%, the number of cleaning cycles could be decreased by up to 11%. Finally, the coated material showed negligible degradation, not exceeding the degradation detected for the reference material.

Keywords: concentrated solar thermal power; water saving; solar reflector; anti-soiling coating; outdoor test; reflectance measurement; soiling rate; cleaning method

---

1. Introduction

Concentrated solar thermal power (CSP) plants are integrated by large reflector surfaces that concentrate the direct normal irradiation (DNI) onto a receiver, where available solar energy is converted to useful thermal energy [1]. To avoid optical losses in the energy conversion process, it is crucial that the concentrating solar reflectors are kept as clean as possible, because soiling accumulated on them reduces their reflectance and consequently, the plants efficiency [2,3]. Cleaning the solar field implies high operation and maintenance (O&M) costs [4] and also creates an important issue in areas with water scarcity, which normally match with high DNI availability zones, where CSP plants are typically located. Consequently, a reduction of the soiling rate of the solar field and the cleaning needs, is one of the main challenges of this technology at present.

Among the different technical solutions implemented or under study in order to mitigate the soiling, special coatings deposited on the reflector's front surface (typically named as "anti-soiling", "self-cleaning" or "easy-to-clean" coatings) are becoming an attractive approach [5–7]. This technology is used in a variety of products nowadays, with glazing products being the foremost area of application [8]. Solar energy applications are also taking advantage of the efficiency increase over time, due to the soiling reduction in the optical surfaces through this kind of coatings [9], mainly in photovoltaic technology [10–12] but also more recently in CSP plants [13–15]. For CSP applications, anti-soiling coatings are applied both to the glass tubes of parabolic-trough collectors and to solar reflectors of all different concentrators. In this case, the advantage is expected not only with the increase of the plant's efficiency, but also in the reduction of the water consumption and O&M costs, thanks to having to carry out less cleaning activities.

IK4-Tekniker and Rioglass Solar are developing an innovative anti-soiling coating for concentrating solar reflectors [16,17]. In this case, the product is based on the hydrophilic effect, which consists on high surface energy and low contact angles between the surface and the water, in order to assure dirt particles' transportation off the surface. To be competitive, a proper anti-soiling coating for reflectors must meet following three requirements:

- Negligible reduction of the reflectance properties in the initial status.
- Appropriate durability over time, which means that the coating must keep its optical properties after being exposed to the weather agents (abrasion, temperature, humidity, pollutants and radiation).
- Appropriate behavior in reducing the dust accumulation on the reflector surface under real outdoor conditions.

While the first two features of the anti-soiling coating developed by IK4-Tekniker and Rioglass Solar were addressed in [18], this paper is focused on studying the behavior of the coating with regard to the last requirements, presenting the results of a long-term outdoor test campaign performed by DLR and CIEMAT at the Plataforma Solar de Almería (PSA).

2. Materials and Methods

This section describes the anti-soiling coating material analyzed in this work, the different cleaning methods used to wash the samples, the equipment used over the test campaign, the measurement frequency and the optical parameters studied.

Six 4 mm silvered-glass reflector samples with dimensions of 20 cm × 40 cm, developed by IK4-Tekniker and Rioglass, were installed at the PSA from 25 October, 2017 until 31 August, 2019 comprising a period of nearly two years. These samples are divided into two parts, one of which is covered with a hydrophilic anti-soiling coating while the other one does not have the coating layer. The reflectors were exposed on two metallic racks with a tilt of 45 degrees in south direction (see Figure 1).

Figure 1. Exposure racks with mirror samples installed in south direction at the Plataforma Solar de Almería (PSA) facilities.

Two different cleaning methods were applied, demineralized pressurized water and a wet brush. For the former, a pressure washer model HDS 10/20-4M manufactured by Kärcher (Winnenden, Germany) was utilized to clean the mirrors situated on the structure S1. The cleaning operations were performed at ambient temperature and 100 bar pressure. In addition, the distance between water pressure nozzle (aperture angle 25°, diameter 0.54 mm) and the samples was kept constant at 50 cm for all the samples. For the latter, a soft brush made with horse tail hair in combination with flowing water was employed as cleaning method for the 2nd structure S2. For both methods, 5 sweeps were done to wash the samples. In order to avoid differences in the methodology, the reflectors were always cleaned by the same operator and in the same conditions. Then, a suitable comparison and evaluation of the cleaning methods can be performed for this anti-soiling treatment. In addition to the regular cleaning, two special cleanings were performed. These resets just comprised the exhaustive cleaning with the brush technique until all soiling was removed from the samples on all structures. The first reset was performed after 8 months and the second after 20 months of exposure. After completing the exposure campaign, the samples returned to the laboratory and their reflectance was measured after abundant cleaning with soft paper tissue and demineralized water.

The optical characterization device used for this study was the 115R-USB reflectometer manufactured by Devices and Services Co. (D&S, Dallas, TX, USA) [19]. This instrument is able to measure the specular reflectance $\rho_{\lambda,\phi}(\lambda,\theta_i,\phi)$, at a the wavelength of $\lambda = 660$ nm, an incidence angle of $\theta_i = 15°$ and an acceptance angle of $\phi = 12.5$ mrad. Furthermore, varying the position of the support screws, it is possible to adapt the height of the reflectometer to the mirror curvature and to adjust the beam path to obtain a correct measurement. In addition, the D&S is lightweight, portable and the influence of diffuse light is negligible, thus, it is appropriate for outdoor measurements. Three measurements were taken on the coated and non-coated surface with a mask to measure always in the same point, where coated part is on the left side and uncoated part is on the right side (see Figure 1). The measurements were taken according to the current version of the SolarPACES Reflectance Guideline [20].

Regarding the measurement frequency, both structures were measured before cleaning every 2 weeks. Structure 1 as well as structure 2 were then washed and measured after cleaning every 2 weeks.

From the reflectance measurements, several meaningful parameters were determined to evaluate the coating. The cleanliness factor, ξ, is defined as the ratio of the actual reflectance, $\rho_{s,\phi}$, to the initial value in the perfectly clean state, $\rho_{s,\phi,\text{clean}}$:

$$\xi = \frac{\rho_{s,\phi}}{\rho_{s,\phi,\text{clean}}} \tag{1}$$

During the campaign, the cleanliness value may be influenced by degradation. The cleanliness value equals 1 at the beginning of the campaign after abundant cleaning in the laboratory with demineralized water and soft paper tissues, followed by removal of remaining particles with pressurized air, to assure maximum reflectance. While the samples stay outside during exposure, the cleaning is usually not able to completely remove all the soiling completely and the reflectance in the perfectly clean state is unknown. If degradation decreases the reflectance in addition to the soiling, this will directly influence the cleanliness parameter.

One important parameter derived from the cleanliness, is the difference between the cleanliness of the coated anti-soiling samples, ξ_{AS}, and the uncoated reference material, ξ_{uncoated}, in the following referred to as cleanliness gain, Δ_ξ:

$$\Delta_\xi = \xi_{AS} - \xi_{\text{uncoated}} \tag{2}$$

In the case that this parameter is positive, the cleanliness of the anti-soiling material is higher than for the uncoated reference material, which means there is an advantage of the anti-soiling coating. Negative Δ_ξ values show a disadvantageous behavior of the coating compared to the reference material.

The accumulated cleanliness gain, $\overline{\Delta_\xi}(t)$, after exposure at a certain exposure time, t, can be calculated by integrating the reflectance difference over the analyzed exposure time dividing by the exposure duration Δt:

$$\overline{\Delta_\xi}(t) = \frac{\int_{t=0}^{t} \Delta_\xi dt}{\Delta t} \tag{3}$$

It represents the mean cleanliness difference, taking into account the varying measurement frequency throughout the campaign. The accumulated cleanliness gain determines the actual advantage of the respective coating compared to the reference material until the analyzed point in time. For the calculation of this value, a linear behavior of the cleanliness between two measurements is assumed.

Finally, one further parameter can be determined supposing this linear behavior, the soiling rate $\hat{\xi}$. The soiling rate is just the change of cleanliness over time and it is usually expressed in pp/day. Within this investigation it is calculated taking the actual cleanliness value before cleaning subtracting the value after cleaning from the last measurement, divided by the time interval between the measurements Δt_c.

$$\hat{\xi}(t) = \frac{\xi_{before}(t) - \xi_{after}(t - \Delta t_c)}{\Delta t_c} \tag{4}$$

3. Results

As the whole outdoor exposure campaign produced a lot of data, these are separated into several groups. Data are separated between the values before cleaning, which means in the soiled state, and the values after cleaning.

In Figure 2 the data for S1 and S2 before cleaning is displayed. Two graphs are included, the first showing the cleanliness factor of the different materials, ξ, together with the amount of daily rain (in mm/day). The second shows the cleanliness gain of the coatings compared to the uncoated material, Δ_ξ.

The following observations can be made from the analysis of these graphs:

- The overall cleanliness factor of samples for S2 is higher than for S1. This is due to the fact that the cleaning with brush is more effective and able to restore the cleanliness basically to its initial values. This fact will be confirmed by the values in the clean state in the following section.
- The cleanliness is higher in periods with a high quantity of rain. Rain usually acts as a natural cleaning mechanism and prevents the samples from excessive soiling.
- The cleanliness gain of the coated samples for both structures is higher when the soiling is stronger and thus the overall cleanliness is lower. The stronger the soiling is, the better the coatings can fulfill their purpose of avoiding the soiling (this can be extracted from the combinations of both graphs).
- The cleanliness gain remains positive for the whole exposure campaign, proving a positive effect of the coatings for all occurring soiling situations. There is only one single event where the coating shows a negative value for S2. This was an event of exceptionally strong soiling due to light rain in combination with a dusty atmosphere (mean cleanliness of 0.66). These events lead to a very inhomogeneous soiling pattern on the samples (Figure 3) and to a high standard deviation of the measurements. With these high deviations, the probability of the occurrence of aberrations is increased. The rain event in this case was too weak to contribute a sufficient quantity to be detected by the rain gauge measurements.
- The highest cleanliness gain Δ_ξ detected was as high as 7.2 pp for S2 after another medium to strong soiling event (mean cleanliness 0.88 on that day).
- The mean cleanliness gain before cleaning is 1.4 pp for S1 and 0.9 pp for S2. When only situations of stronger soiling are evaluated, the mean cleanliness gain is even higher. Looking only at events with a high soiling level (cleanliness below 0.9), the mean cleanliness gain rises up to 2.4 pp (S1) and 1.5 pp (S2).

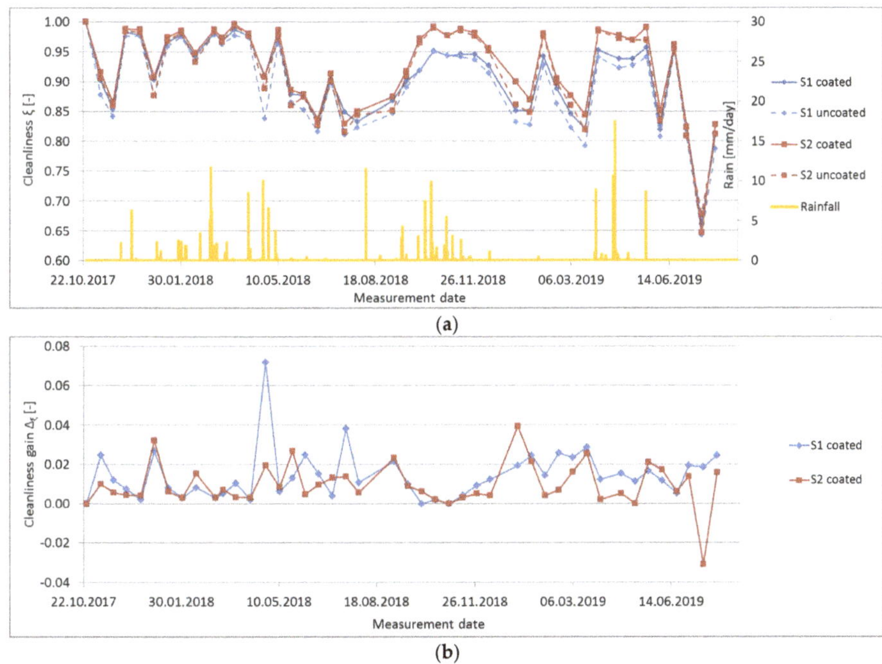

(a)

(b)

Figure 2. Data before cleaning, (**a**): cleanliness factor, (**b**): cleanliness gain.

Figure 3. Sample surface after a strong soiling event, showing an inhomogeneous soiling pattern.

In Figure 4 the data for the values after cleaning is presented, in the same way as in Figure 2 for the before cleaning values. This representation is especially useful to analyze the "easy-to-clean" properties of the coatings as well as the cleaning procedures themselves. Observations that can be drawn from these graphs are the following:

- The cleanliness after cleaning remains higher for S2 during the whole exposure campaign and always reaches values around 1. The values even reach values slightly higher than 1 due to the measurement uncertainty of the reflectometer. Therefore, the cleaning method using the brush is much more effective than the high pressure water method.

- The cleaning method with pressurized water is not able to restore the initial cleanliness values over the whole exposure campaign ($\xi < 1$). Remaining soiling stays on sample surfaces decreasing the reflectance of the samples on S1. The effect is cumulative, showing decreasing values after cleaning in certain periods, especially after the first reset when strong soiling appears (compare Figure 2). This tendency is more pronounced for the uncoated reference material. That also influences the before cleaning values in Figure 2, where the difference between S1 and S2 is especially strong in periods of lower soiling after the accumulation of soiling on S1.
- After the two resets (extensive cleaning, marked in the chart of the cleanliness factor), the initial cleanliness is also nearly restored for S1, indicating negligible degradation.
- The cleanliness gain Δ_ξ after cleaning is lower than before cleaning. This underlines the fact that Δ_ξ is more pronounced with stronger soiling. Nevertheless, it stays in the positive regime for the whole campaign and thus proves the advantage of the anti-soiling coatings and indicates the positive effect on the "easy-to-clean" property.
- The mean cleanliness gain is also higher for S2 because of the lower cleanliness of S1.

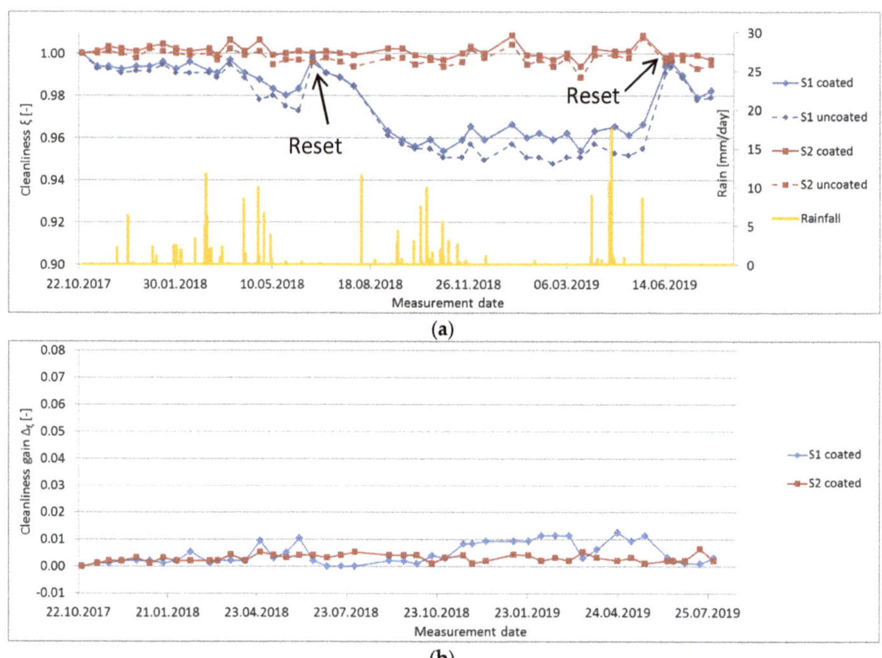

Figure 4. Data after cleaning, (**a**): cleanliness factor, (**b**): cleanliness gain.

To evaluate if degradation affected the samples during outdoor exposure, their reflectance was measured in the laboratory before and after the exposure. In Table 1 the initial and final ξ values are displayed together with the resulting cleanliness loss, $\Delta\xi$. The measured losses do not exceed 0.004 in any of the cases, which is still within the uncertainty of the reflectometer. This leads to the conclusion that no perceivable degradation affects the materials within the tested duration.

Table 1. Reflectance measurements performed in the laboratory before and after exposure together with the reflectance loss.

	Material	Initial ξ [–]	Final ξ [–]	Cleanliness Loss Due to Degradation, Δξ [–]
S1	Coated	1.000	0.997 ± 0.006	−0.003 ± 0.006
	Uncoated	1.000	0.996 ± 0.003	−0.004 ± 0.003
S2	Coated	1.000	1.001 ± 0.003	0.001 ± 0.005
	Uncoated	1.000	0.999 ± 0.002	−0.001 ± 0.004

In Figure 5 the accumulated cleanliness gain, $\overline{\Delta_\xi}(t)$, is displayed. In the beginning fluctuations of the values are stronger, because the overall duration is lower. The values become more stable with longer exposure. $\overline{\Delta_\xi}(t)$ remains higher for S1 for the whole exposure period. The final values of the curves after 22 months of exposure also represent the overall accumulated cleanliness gain of the coatings regarding the whole campaign. These values are $\overline{\Delta_{\xi,S1}}$ (22 months) = 1.0 pp and $\overline{\Delta_{\xi,S2}}$ (22 months) = 0.7 pp for the two structures. The better behavior of the coated samples exposed in the structure S1 is due to the higher soiling level of these samples during the outdoor exposure, as a consequence of the lower efficiency of the cleaning method applied. This fact indicates that the anti-soiling coating developed will show a better behavior in areas with higher soiling rates, such as the arid zones where the CSP plants are installed.

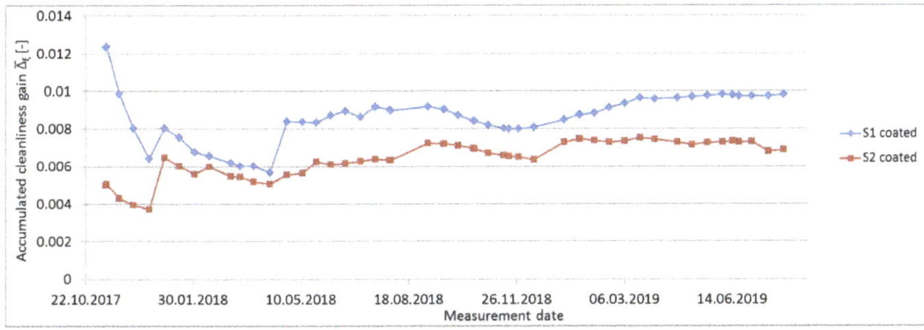

Figure 5. Accumulated cleanliness gain for both structures.

The development of the soiling rate over the exposure period is displayed in Figure 6. The values are always negative as they present a loss in cleanliness over time. Higher soiling rates can be appreciated for the uncoated material. As already seen in the graphs for the cleanliness factor before cleaning, soiling rates are lower for rainy periods. Soiling rates reach values of up to 2.3 pp/day in cases of strong soiling.

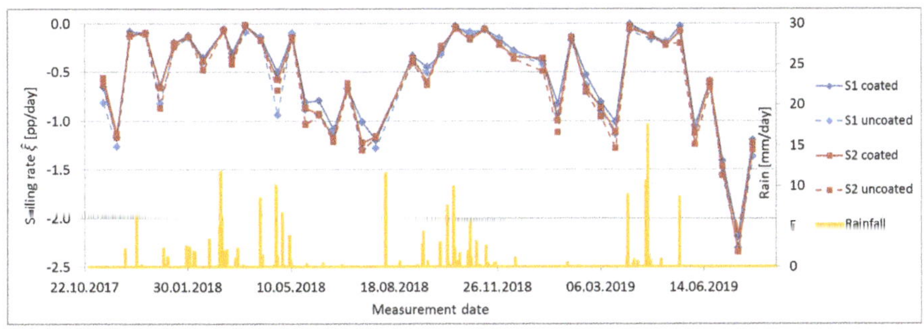

Figure 6. Soiling rate for both materials and structures.

In Table 2 the mean soiling rates are displayed for coated and uncoated samples on both structures. The soiling rate for the coated material is lower than for the uncoated material (average of both structures is 0.54 instead of 0.60 pp/day), once again proving the positive effect of the anti-soiling coating.

Table 2. Mean soiling rate, time until cleaning and number of cleanings per year.

Material	S1 Coated	S1 Uncoated	S2 Coated	S2 Uncoated
Average soiling rate [pp/day]	−0.52	−0.59	−0.56	−0.60
Time until cleaning [days]	7.72	6.81	7.16	6.63
Number of cleanings per year	47.34	53.63	50.99	55.06

The Wascop project, during which this current study has been conducted, is focused on the water saving potential of different technologies. One possibility to reduce water consumption by the use of anti-soiling coatings is to decrease the cleaning frequency. To calculate the benefit of the coating, a fixed cleanliness threshold of $\xi = 0.96$ is assumed, below which cleaning is performed. Taking into account the mean soiling rates presented above, the duration when this value is reached can be calculated. The number of days until cleaning is as well given in Table 2. The longer intervals between cleaning events would therefore result in 6.3 performed events less per year for S1, and respectively 4.1 events less for S2, when using the anti-soiling coating. This is a reduction of the number of performed cleanings of 11.7% and 7.4% respectively. It is important to highlight that this calculation has been performed for a specific site and under several assumptions, such as a constant soiling rate, a constant cleaning frequency every two weeks and a fixed cleanliness threshold of 0.96. Therefore, the results obtained are only illustrative.

4. Conclusions

The results of the herein presented outdoor exposure campaign of a novel anti-soiling coating for silvered-glass solar reflectors, shows the benefit of the application of these coatings compared to state-of-the-art commercial silvered-glass mirrors. The main benefits of the coating are the following:

- The anti-soiling effect of the applied coatings leads to a higher cleanliness of the reflectors throughout the exposure campaign. Taking into account the whole campaign, an accumulated cleanliness gain of 1.0 and 0.7 pp is reached, for pressurized water and brush cleaning respectively. When only the samples in the soiled state (before cleaning) are evaluated, the accumulated cleanliness gain increases to 1.4 and 0.9 pp, respectively. These values are even higher for situations of stronger soiling: for a cleanliness below 90%, the accumulated cleanliness gain increases to 2.4 and 1.5 pp respectively. The detected maximum momentary cleanliness gain reaches over 7 pp for a single soiling event.
- Lower soiling rates are detected for the anti-soiling coated samples, with a minimum value of 0.52 pp/day for the coated samples on S1, and a maximum of 0.60 pp/day for the uncoated samples on S2.
- The "easy-to-clean" properties of the anti-soiling coatings facilitate the cleaning process and thus help to recover the initial cleanliness of the reflectors.
- Fewer cleaning cycles have to be performed for the coated mirrors to reach the same mean cleanliness compared to uncoated mirrors. For a constant threshold cleanliness of 0.96, the number of cleaning cycles in Tabernas, Spain can be reduced by 7% to 12% for brush cleaning and pressurized water, respectively, considering a constant soiling rate and a fixed cleaning frequency of 2 weeks.
- The coatings show excellent durability during the course of the whole campaign and showing no signs of degradation.

- The investigated site represents an environment with relatively little dust in the atmosphere, leading to low soiling rates. The performance of the coatings is expected to be better for high soiling sites because the coating has demonstrated a higher effectiveness with stronger soiling.

The investigated coatings are still in the development phase. For a final evaluation, the commercial viability of the application of the coatings has to be studied before implementation. The result of this study will be basically the balance of the additional cost of the coating application and the expected benefit for the plant operation. The cost of the coating will mainly depend on the size of the solar field and the type of reflector to be installed. The benefit of the coatings is a complex subject which depends on many factors, among others plant design and operation (size of solar field and storage, technology used, operational strategy, cleaning technique) as well as the location (environmental conditions, soiling development, water availability).

Author Contributions: Conceptualization, F.S., A.F.-G., I.A. and G.P.; methodology, A.F.-G. and F.S.; software, J.W., D.A.-A. and F.B.-M.; validation, F.S. and A.F.-G.; formal analysis, J.W.; investigation, J.W., D.A.-A. and F.B.-M.; resources, F.S. and A.F.-G.; data curation, J.W.; writing—original draft preparation, J.W.; writing—review and editing, J.W., F.S., A.F.-G., D.A.-A., F.B.-M., I.A. and G.P.; visualization, J.W.; supervision, F.S., A.F.-G., I.A. and G.P.; project administration, F.S. and A.F.-G.; funding acquisition, F.S., A.F.-G., I.A. and G.P.

Funding: This research has received funding from the European Union's Horizon 2020 research and innovation program under grant agreement No 654479 (Wascop).

Acknowledgments: The authors want to thank Tomás Reche Navarro and Carmen Amador Cortés from DLR and Lucía Martinez Arcos from CIEMAT for their valuable contributions to the realization of this work, especially on the exposure campaign and measurements.

Conflicts of Interest: The authors declare no conflict of interest.

References

1. Mills, D. Advances in solar thermal electricity technology. *Sol. Energy* **2004**, *76*, 13–19.

2. Bouaddi, S.; Fernández-García, A.; Ihlal, A.; Ait El Cadia, R.; Álvarez-Rodrigo, L. Modeling and simulation of the soiling dynamics of frequently cleaned reflectors in CSP plants. *Sol. Energy* **2018**, *166*, 422–431.

3. Wolfertstetter, F.; Wilbert, S.; Dersch, J.; Dieckmann, S.; Pitz-Paal, R.; Ghennioui, A. Integration of soiling-rate measurements and cleaning strategies in yield analysis of parabolic trough plants. *J. Sol. Energy Eng.* **2018**, *140*, 041008–041011. [CrossRef]

4. Bouaddi, S.; Fernández-García, A.; Sansom, C.; Sarasua, J.A.; Wolfertstetter, F.; Bouzekri, H.; Sutter, F.; Azpitarte, I. A review of conventional and innovative-sustainable methods for cleaning reflectors in concentrating solar power plants. *Sustainability* **2018**, *10*, 3937. [CrossRef]

5. Sarver, T.; Al-Qaraghuli, A.; Kazmerski, L. A comprehensive review of the impact of dust on the use of solar energy: History, investigations, results, literature, and mitigation approaches. *Renew. Sustain. Energy Rev.* **2013**, *22*, 698–733. [CrossRef]

6. Costa, S.C.S.; Diniz, A.S.A.C.; Kazmerski, L.L. Dust and soiling issues and impacts relating to solar energy systems: Literature review update for 2012–2015. *Renew. Sustain. Energy Rev.* **2016**, *63*, 33–61. [CrossRef]

7. Atkinson, C.; Sansom, C.L.; Almond, H.J.; Shaw, C.P. Coatings for concentrating solar systems—A review. *Renew. Sustain. Energy Rev.* **2015**, *45*, 113–122. [CrossRef]

8. Midtdal, K.; Jelle, B.P. Self-cleaning glazing products: A state-of-the-art review and future research pathways. *Sol. Energy Mater. Sol. Cells* **2013**, *109*, 126–141. [CrossRef]

9. Polizos, G.; Sharma, J.K.; Smith, D.B.; Tuncer, E.; Park, J.; Voylov, D.; Sokolov, A.P.; Meyer, H.M.; Aman, M. Anti-soiling and highly transparent coatings with multi-scale features. *Sol. Energy Mater. Sol. Cells* **2018**, *188*, 255–262. [CrossRef]

10. Piliougine, M.; Cañete, C.; Moreno, R.; Carretero, J.; Hirose, J.; Ogawa, S.; Sidrachde-Cardona, M. Comparative analysis of energy produced by photovoltaic modules with anti-soiling coated surface in arid climates. *Appl. Energy* **2013**, *112*, 626–634. [CrossRef]

11. Syafiq, A.; Pandey, A.K.; Adzman, N.N.; Rahim, N.A. Advances in approaches and methods for self-cleaning of solar photovoltaic panels. *Sol. Energy* **2018**, *162*, 597–619. [CrossRef]

12. Moraes Lopes de Jesus, M.A.; Timò, G.; Agustín-Sáenz, C.; Braceras, I.; Cornelli, M.; de Mello Ferreira, A. Anti-soiling coatings for solar cell cover glass: Climate and surface properties influence. *Sol. Energy Mater. Sol. Cells* **2018**, *185*, 517–523. [CrossRef]

13. Schwarberg, F.; Schiller, M. Enhanced solar mirrors with anti-soiling coating. In Proceedings of the 18th Int Conference on CSP and Chemical Energy Systems, Marrakech, Morocco, 11–14 September 2012.

14. Polizos, G.; Schaeffer, D.A.; Smith, D.B.; Lee, D.F.; Datskos, P.G.; Hunter, S.R. Enhanced durability transparent superhydrophobic anti-soiling coatings for CSP applications. In Proceedings of the ASME 8th Int Conference on Energy Sustainability, Boston, MA, USA, 30 June–2 July 2014.

15. Plesniak, A.P.; Pfefferkorn, C.; Hunter, S.R.; Smith, D.B.; Polizos, G.; Schaeffer, D.A.; Lee, D.F.; Datskos, P.G. Low cost anti-soiling coatings for CSP collector mirrors and heliostats. In Proceedings of the SPIE 9175, High and Low Concentrator Systems for Solar Energy Applications IX, San Diego, CA, USA, 7 October 2014.

16. Ubach, J.; Gómez, E.; Zarrabe, H.; Aranzabe, E. Coated Glass for Solar Reflectors. U.S. Patent EP 3090990 A1, 4 May 2015.

17. Aranzabe, E.; Azpitarte, I.; Fernández-García, A.; Argüelles, D.; Pérez, G.; Ubach, J.; Sutter, F. Hydrophilic anti-soiling coating for improved efficiency of solar reflectors. In Proceedings of the AIP Conference Proceedings, Santiago de Chile, Chile, 26–29 September 2018; Volume 2033, p. 220001.

18. Fernández-García, A.; Aranzabe, E.; Azpitarte, I.; Sutter, F.; Martínez-Arcos, L.; Reche-Navarro, T.J.; Pérez, G.; Ubach, J. Durability testing of a newly developed hydrophilic anti-soiling coating for solar reflectors. In Proceedings of the 24th Solar PACES International Conference on CSP and Chemical Energy Systems, Casablanca, Morocco, 2–5 October 2018.

19. Fernández-García, A.; Sutter, F.; Martínez-Arcos, L.; Sansom, C.; Wolfertstetter, F.; Delord, C. Equipment and methods for measuring reflectance of concentrating solar reflector materials. *Sol. Energy Mater. Sol. Cells* **2017**, *167*, 28–52. [CrossRef]

20. Fernández-García, A.; Sutter, F.; Montecchi, M.; Sallaberry, F.; Heimsath, A.; Heras, C.; Le Baron, E.; Soum-Glaude, A. *Parameters and Method to Evaluate Reflectance Properties of Reflector Materials for Concentrating Solar Power Technology—Official Reflectance Guideline Version 3.0*; SolarPACES: Tabernas, Spain, March 2018.

Article

Scale Formation and Degradation of Diffusion Coatings Deposited on 9% Cr Steel in Molten Solar Salt

Ceyhun Oskay *, Tobias M. Meißner, Carmen Dobler, Benjamin Grégoire and Mathias C. Galetz

DECHEMA-Forschungsinstitut, Theodor-Heuss-Allee 25 60486 Frankfurt (Main), Germany; meissner@dechema.de (T.M.M.); carmen.dobler@vtu.com (C.D.); benjamin.gregoire@dechema.de (B.G.); mathias.galetz@dechema.de (M.C.G.)
* Correspondence: ceyhun.oskay@dechema.de

Received: 27 September 2019; Accepted: 18 October 2019; Published: 22 October 2019

Abstract: The employment of ferritic-martensitic steels e.g., P91, as structural materials in concentrated solar power (CSP) plants can significantly increase cost-efficiency. However, their application is strongly restricted by their lower corrosion resistance in molten nitrates, compared to austenitic steels or Ni-based alloys. In this study, Cr-, Al-, and Cr/Al-diffusion coatings were deposited on P91 via pack cementation in order to improve its scaling behavior in molten solar salt (MSS). The corrosion behavior of coated specimens was investigated with respect to uncoated P91 in MSS at 600 °C for up to 1000 h. The exposure in MSS resulted in a thick, highly porous, and multi-layered oxide scale on uncoated P91 consisting of hematite, magnetite, and sodium ferrite. On the other hand, the scale grown on the chromized P91 comprised of a thin Cr-rich inner layer, which shifted breakaway to prolonged exposure durations. The aluminized specimens both formed very thin, highly protective alumina scales with localized protrusions.

Keywords: concentrated solar power; grade 91 steel; Cr-diffusion coating; Al-diffusion coating; pack cementation; molten nitrate corrosion; X-ray diffraction; Raman spectroscopy; third element effect

---

1. Introduction

With the ultimate goal of reducing CO_2 emissions, solar energy has been qualified as one of the most prolific green energy technologies (owing to the abundance of sunlight and the "climate neutral" nature of the related energy conversion methods). Along with photovoltaics (PV), concentrated solar power (CSP) technology has received considerable interest in terms of research and development activities, leading to its commercialization over the past decades [1,2]. CSP technology utilizes programmable heliostats, which concentrate solar irradiation to a so-called receiver. The focused solar energy is then converted to thermal energy and subsequently to electricity via a conventional steam turbine [3–5].

CSP plants can be operated with or without thermal energy storage (TES) systems; the former enabling improved dispatchability for electricity generation. Hence, the current development trend focuses on CSP plants with TES (e.g., Crescent Dunes Solar Power Plant, USA) by the utilization of heat transfer fluids (HTF) with high thermal capacities. Molten nitrate salts, particularly the non-eutectic mixture of 60 wt.% $NaNO_3$-40 wt.% KNO_3 (known as "solar salt") are qualified as state-of-the-art HTF for CSP applications owing to their beneficial thermophysical properties for TES [1,4,6–11]. The minimum temperature of the operation range of solar salt (around 290 °C) is limited by the liquidus temperature (lying around 240 °C) to avoid solidification of the salt, since continuous molten salt flow is required [2,12]. On the other hand, the maximum temperature of the operation range is restricted by the thermal decomposition of nitrate initially to nitrite (Reaction 1) and further to nitrous oxides (Reaction 2), which is accelerated at temperatures exceeding 600 °C [1,8,10,13].

$$NO_3^- \rightleftharpoons NO_2^- + \frac{1}{2}O_2(g) \tag{1}$$

$$2NO_2^- \rightleftharpoons NO_2(g) + NO(g) + O^{2-} \tag{2}$$

These decomposition reactions alter the salt composition by changing the nitrate/nitrite ratio and thus affect the thermophysical properties, as studied in [14]. More importantly, both decomposition reactions result in an increased concentration of oxidizing agents in the salt melt, which is basically an ionic electrolyte at higher temperatures and conveys oxidizing agents to metallic surfaces, while metal cations can be dissolved into the salt melt [2,10,13]. For instance, Cr is a stable oxide former for various industrial applications; however, Cr cations present in the Cr-rich oxide scales can further oxidize and form chromate or dichromate species (e.g., via Reaction 3 [7,15]) that are highly soluble in the molten salt [9,11,16]. The formation of these species not only increases the extent of Cr-depletion, but also generate major environmental and health issues due to chromates being highly toxic [6].

$$Cr_2O_3 + 2O^{2-} + \frac{3}{2}O_2(g) \rightleftharpoons 2CrO_4{}^{2-} \tag{3}$$

Furthermore, commercial solar salt contains a wide variety of impurities, such as chlorides and sulfates, which can increase the corrosion rate, as shown in [17,18]. Due to the above stated reasons, the corrosion resistance of structural materials employed in the receiver tubing system and the hot storage tank is a decisive factor for plant design and lifetime [4,6,9–11,19].

Another important obstacle currently restricting the wider application of the CSP technology is its lower cost-efficiency with respect to other renewable energy sources e.g., PV, hydro- or wind-power [5]. Employment of cheaper alloys as structural materials can be a very efficient method to reduce costs in CSP plants. The corrosion behavior of low Cr-steels, austenitic steels and Ni-based alloys in molten nitrate salts at a wide temperature range has been investigated in different studies [2,10,13,14,17,18,20–22], which showed higher corrosion resistance for austenitic steels and Ni-based alloys compared to ferritic-martensitic steels. Notwithstanding their high corrosion resistance, employment of austenitic steels or Ni-based alloys significantly increases the cost of structural materials in CSP plants. For instance, the unit price of the Ni-based alloy 617 is roughly nine times higher than that of the grade 91 steel [23]. Corrosion resistance of ferritic-martensitic steels in molten nitrate salt can be increased via diffusion coatings, thereby enabling their employment in the receiver panels. Recent studies showed that aluminide coatings can significantly increase the corrosion resistance of ferritic-martensitic steels in molten nitrate salts by forming Al-rich protective oxide scales [6,12,24–26]. This study aims at the elucidation of corrosion and protection mechanisms by identifying the chemical composition of oxide scales grown on uncoated and diffusion coated ferritic-martensitic steels in molten solar salt (MSS) using complementary characterization methods. For this purpose, Cr-, Al- and Cr/Al-diffusion coatings were manufactured on grade 91 steel, which is widely employed in heat exchanger systems up to 650 °C [27]. Subsequently, static immersion tests in MSS were conducted at 600 °C. The scaling behavior of coated specimens was investigated and compared to that of uncoated P91 to explore the applicability of diffusion-coated ferritic-martensitic steels as structural materials in CSP systems with TES.

2. Materials and Methods

Ferritic-martensitic steel X10CrMoVNb9-1 (P91/T91) was used as a base material and coated according to the description in the following subsection. Its chemical composition is shown in Table 1:

Table 1. Nominal composition (in wt.%) of the ferritic-martensitic steel, P91, used in this study [27].

Alloy	Fe	C	Cr	Ni	Mn	Si	P	S	Mo	V	Al	Nb	N
P91	bal.	0.08–0.12	8–9.5	≤0.4	0.3–0.6	0.2–0.5	≤0.02	≤0.01	0.85–1.05	0.18–0.25	≤0.04	0.06–0.1	0.03–0.07

Coupon specimens of dimensions 20 mm × 10 mm × 3 mm were manufactured using wire cutting and cleansed of native oxides and contamination by means of glass-bead blasting and ultrasonic degreasing in ethanol. All test samples were weighed after every coating step and prior to exposure, respectively. Additionally, their dimensions were measured followed by a calculation of each sample's individual surface area to finally assess specific weight changes.

2.1. Coating Manufacturing

All coatings were manufactured via the well-established powder pack cementation chemical vapor deposition (CVD) process, which is described in detail elsewhere e.g., in [28–30]. Cr- and Al-diffusion coatings were manufactured via single-step processes using the parameters summarized in Table 2, respectively. Cr/Al-diffusion coatings were produced by combining both as a two-step process, similar to the procedure used in [31].

Table 2. Overview of the manufacturing parameters for Cr- and Al-coatings as well as the Cr/Al-coating for which both single-step processes were carried out successively.

Diffusion Element	Powder Composition			Process Parameters		
	Master Alloy	Activator	Inert Filler	Temperature	Time	Atmosphere
Cr	Cr	$MnCl_2$	Al_2O_3	1050 °C	2 h	Ar + 5% H_2
Al	Al	NH_4Cl	Al_2O_3	1000 °C	1 h	Ar + 5% H_2

2.2. Exposure in Molten Solar Salt

For the isothermal exposures in molten nitrate, state-of-the-art solar salt consisting of 60 wt.% $NaNO_3$ and 40 wt.% KNO_3 was used with initial impurities of 131 ppm Cl^- and 60 ppm SO_4^{2-}. Since solar salt with a relatively higher impurity level is lower-priced and thus technically more relevant compared to pure grades, a higher impurity level of 500 ppm for both chloride and sulfate species was investigated in this study. Initially, both components of the solar salt were pre-mixed to yield a 2 kg salt mixture. The targeted impurity level of chloride and sulfate species was then adjusted by adding NaCl and Na_2SO_4 to the pre-mixed solar salt using a precision (0.01 mg) weighing balance (Mettler Toledo XP205, Columbus, OH, USA). The final salt mixture was then homogenized for 2 h at room temperature using a turbular mixer. The corrosion tests were conducted in a horizontal tube furnace with a test atmosphere of flowing dry synthetic air (4 L/h). A schematic representation of the experimental setup is given in Figure 1. The exposures were carried out following the specifications of ISO 17245:2015 [32] by placing the test specimens in individual cylindrical alumina crucibles, covering them with the salt mixture at room temperature, and heating them to the static test temperature of 600 °C. After each time interval (115, 300 and 1000 h), two samples of the uncoated substrate and of every coating were removed from the hot furnace, cooled to room temperature, and carefully cleaned in warm distilled water.

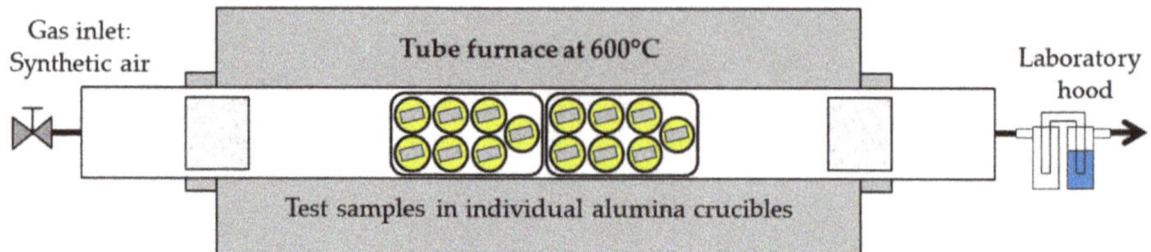

Figure 1. Schematic illustration of the corrosion test rig for exposure in molten solar salt at 600 °C.

2.3. Post-Exposure Characterization

Prior to metallographic preparation, all samples were weighed using a precision weighing balance (Mettler Toledo XP205, Columbus, OH, USA). Phase analysis of the test pieces was performed using a Bruker D8 X-ray diffractometer (Billerica, MA, USA) with Cu-K-alpha radiation and a Lynxeye-semiconductor detector.

In order to enhance the contrast between corrosion or oxidation products and the adjacent epoxy resin, the samples were electroplated with Ni prior to mounting in hot resin. Afterwards, cross-sections of all test specimens were prepared by conventional metallographic methods, including grinding with successively finer SiC paper up to 1200 grit and polishing with 3 and 1 μm diamond suspensions. Inspection of the cross sections was carried out with the help of standard micro-analytical methods such as optical light microscopy and complementary scanning electron microscopy (SEM, Philips XL-40, Amsterdam, The Netherlands) analysis as well as electron probe micro-analysis (EPMA, JEOL JXA-8100, Tokyo, Japan). To further assess the composition of the thin oxide scales grown after exposure, Raman spectroscopy analyses (RENISHAW InVia working with a He-Ne laser, 632.8 nm, Wotton-under-Edge, UK) were also conducted.

3. Results and Discussion

3.1. Microstructure of as-deposited Coatings

3.1.1. Cr-Diffusion Coating

Figure 2 shows the cross-sectional back-scattered electron (BSE) micrograph, elemental distribution maps, and line scans of Fe, Cr, and C, both acquired by EPMA as well as the X-ray diffraction (XRD) pattern of the Cr-diffusion coated P91. The coating consisted of a thin (3–5 μm) $Cr_{23}C_6$ layer (as identified by the XRD analysis shown in Figure 2d) followed by an approximately 30 μm thick Cr-diffusion zone, which contained an average Cr-concentration of 17 at.% (see Figure 2c). The formation of this $Cr_{23}C_6$ layer can be attributed to outward carbon diffusion from the substrate due to the high Cr activity at the surface [33,34]. In addition, Mn was enriched to an average concentration of 1 at.% (corresponds approximately to 1.1 wt.%) in the Cr-diffusion zone (see Figure 2b), compared to its original concentration in P91 (0.3–0.6 wt.%). The enrichment of Mn can be explained by the utilization of $MnCl_2$ as the activator. Mn evidently diffused to the alloy surface to some extent together with Cr. Within the Cr-diffusion zone, large ferrite grains were formed. The stabilization of ferrite is associated with the Cr-enrichment and reduced carbon content due to the formation of the chromium carbide surface layer. Refractory metal (Cr, V, and Mo) rich carbides were observed at the ferrite grain boundaries (see Figure 2b), as also was highlighted by Meier et al. [35].

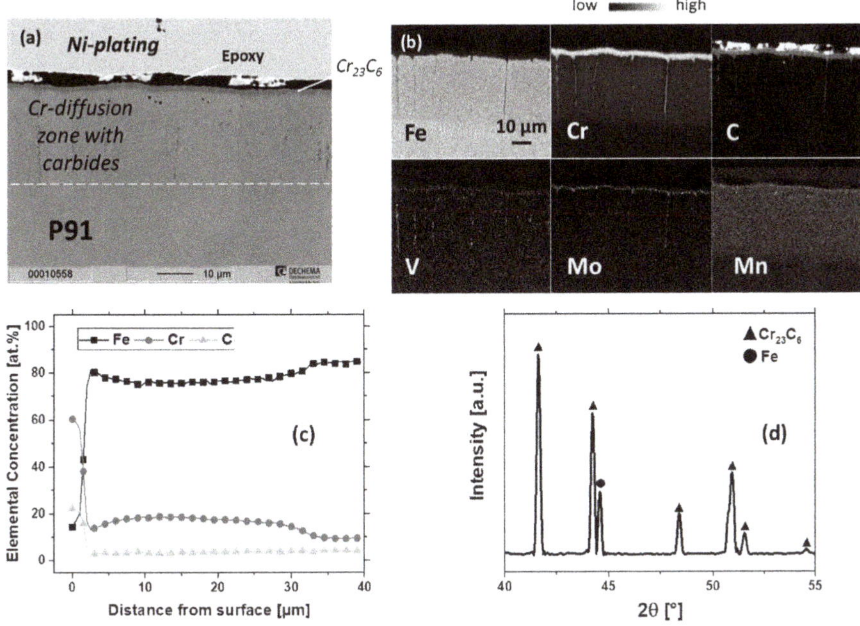

Figure 2. (**a**) Cross-sectional back scattered electron (BSE) image, (**b**) elemental distribution maps and (**c**) quantitative line scans of Fe, Cr and C as well as (**d**) X-ray diffraction (XRD) pattern of Cr-diffusion coated P91. Please note the low adherence of the Ni-plating, resulting in the incorporation of the epoxy into the surface of the coating during mounting.

3.1.2. Al-Diffusion Coating

Pack aluminizing at 1000 °C resulted in the formation of different aluminide layers with varying chemical composition and microstructure (see Figure 3). The Al-concentration at the thin (ca. 5 µm) sub-surface layer was determined to be 65 at.% (Figure 3c). Correspondingly $FeAl_2$ and Fe_2Al_5 phases were identified by the XRD analysis (Figure 3d). Such brittle Al-rich Fe-Al intermetallic phases are usually observed in aluminide coatings manufactured on ferritic steels at lower coating temperatures around 700 °C, provided that pure Al is used as the diffusion element [31,36–38]. On the other hand, aluminizing at elevated temperatures (e.g., at 1000 °C and higher) results in the transformation of the Fe_2Al_5 to FeAl due to the concurrent inward Al and outward Fe diffusion, as further explained in [25,39,40]. Evidently, the selected pack aluminizing parameters in this study involving 1 h dwell at 1000 °C (see Table 2) were not sufficient to transform the Al-rich intermetallic phases to FeAl completely. Together with the insufficient time provided for inward Al diffusion, the continuous Al supply from the pack mixture to the alloy might have resulted in the stabilization of these phases.

The intermediate layer possessed a lower Al-content (nearly 48 at.%) and a significantly higher Fe-content (approximately 45 at.%). Correspondingly this layer comprised of the FeAl phase and exhibited Kirkendall voids (filled with polishing media, see C map in Figure 3b). The formation of such voids can be attributed to the imbalance in the diffusion fluxes of Fe and Al in the counter directions [25,39]. Furthermore, formation and propagation of cracks within the outermost and intermediate layers was observed (Figure 3a). This can be explained by the high discrepancy between the coefficients of thermal expansion (CTE) of Fe-Al intermetallic phases (e.g., CTE of FeAl at 873 K is $21 \cdot 10^{-6}$ K^{-1} [41]) and the alloy (e.g., CTE of P91 at 873 K is $14.1 \cdot 10^{-6}$ K^{-1} [27]) and the consequential build-up of tensile stresses in the coating during the cooling period, as highlighted in [36,37,42].

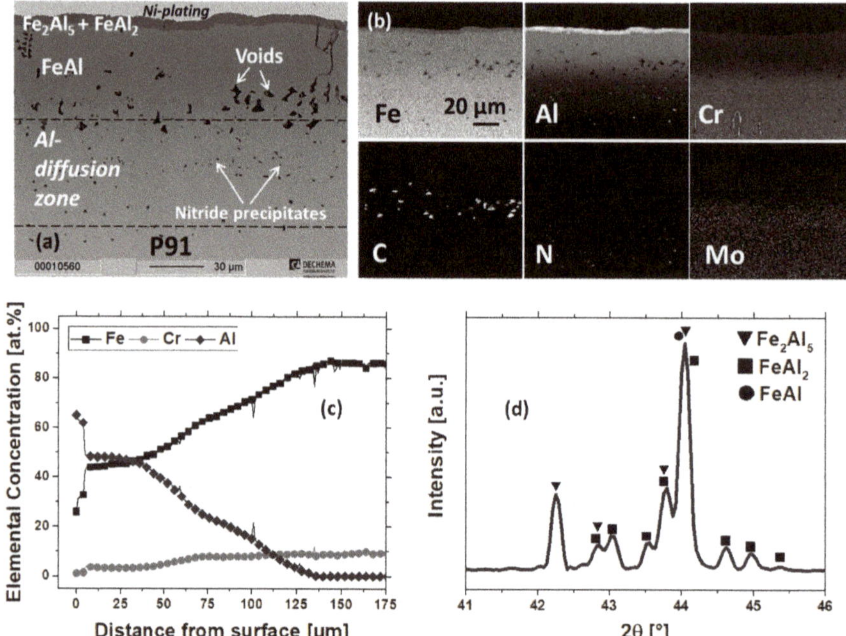

Figure 3. (**a**) Cross-sectional BSE-image, (**b**) elemental distribution maps and (**c**) quantitative line scans of Fe, Cr and Al as well as (**d**) XRD-pattern of aluminized P91.

Underneath the intermediate layer, Al-diffusion into the alloy led to the formation of a ferritic zone in which Al nitride precipitates (see Al and N maps in Figure 3b) were observed. This is in good accordance with [36,39] and can be associated with the low solubility of N in the ferrite zone formed due to the inward diffusion of Al [43]. Zhang et al. showed that the coalescence of such needle-like AlN precipitates at the coating/substrate interface can lead to delamination of the aluminide coating deposited on substrates with a high N-content [43]. Nevertheless, the aluminide coating deposited in this study showed the dispersion of AlN precipitates in the Al-diffusion zone (see N map in Figure 3b) instead of their coalescence at the coating/substrate interface.

3.1.3. Cr/Al-Diffusion Coating

Figure 4 shows the cross-sectional BSE-image, elemental distribution maps, line scans of Fe, Cr and Al, and XRD-pattern of the Cr/Al-coating, which exhibited a similar microstructure and distinctive zones with differing chemical composition to the Al-coating (Figure 3). The $Cr_{23}C_6$ layer present in the as-chromized microstructure (see Figure 2) was dissolved during the aluminizing treatment. Al-diffusion into the former Cr-diffusion zone (Figure 2) led to the formation of an Al-rich (mean concentration of 60 at.%) outer layer with a larger thickness (ca. 40 µm) compared to the pure aluminide coating (see Figure 3; please note the different magnification). Similar to the aluminide coating, the outer layer of the Cr/Al-coating consisted of $FeAl_2$ and Fe_2Al_5 phases, but also contained Cr-rich precipitates (see Cr map in Figure 4b), which were identified as the $(Fe,Cr)_5Al_8$ phase via XRD analysis (Figure 4d). The significant content of Fe determined in the Cr_5Al_8 phase (around 29 at.%) can be explained by the substitution of Cr by Fe in the Cr_5Al_8 phase, as reported by Palm [44]. Unlike Fe, the Cr-profile fluctuated strongly and its lower concentration (see Figure 4c) in the matrix agreed well with its reported maximum solubility of 4.1 at.% in $FeAl_2$ and 6.5 at.% in Fe_2Al_5 [45]. Due to its limited solubility in these Al-rich Fe-Al phases, the excess Cr in the outer layer resulted in the formation of the

Cr_5Al_8 precipitates, in which a significantly higher maximum Cr-concentration of 16 at.% was detected (see Figure 4c). Similar to chromium, the Mo-content increased up to 0.8 at.% (line scan not shown) within these precipitates (see Figure 4b). The enrichment of molybdenum in the Cr_5Al_8 precipitates derives from the refractory metal rich carbides at the grain boundaries as well as in the lower part of the $Cr_{23}C_6$ layer of the as-chromized microstructure (see Mo map in Figure 2b).

The inner layers of the Cr/Al-diffusion coating showed a strong Al-concentration gradient and a very similar microstructure to the Al-diffusion coating. Within the FeAl layer Kirkendall voids, fine nitride precipitates as well as cracks perpendicular to the surface were observed. The formation of cracks can again be explained by the build-up of tensile stresses during the cooling period of coating manufacturing due to the CTE mismatch between the coating phases and the substrate analogous to the aluminized specimen. Underneath the FeAl layer, an almost 60 µm thick Al-diffusion zone with finely dispersed Al nitride precipitates was formed.

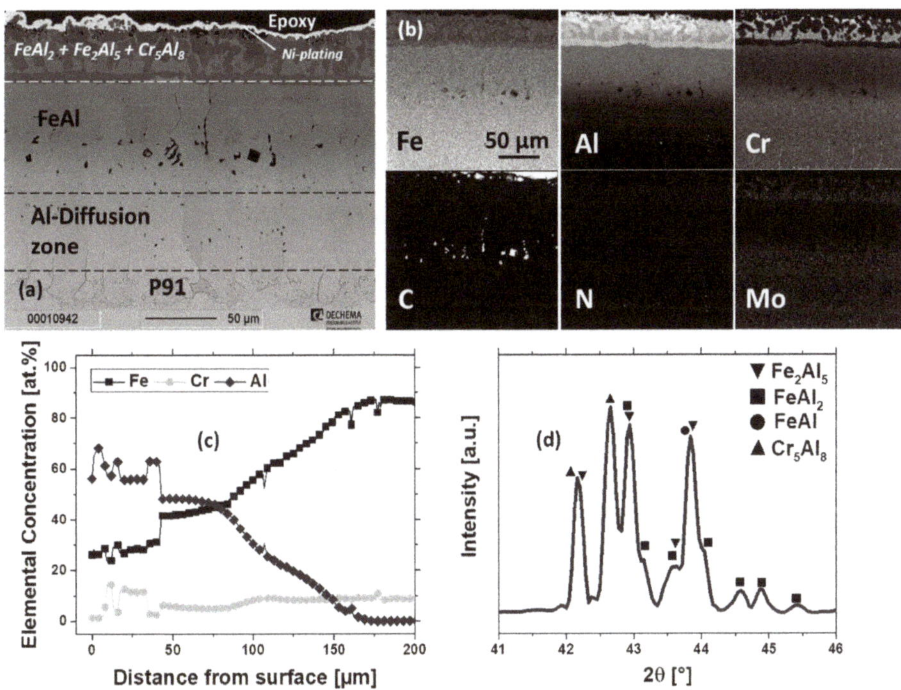

Figure 4. (**a**) Cross-sectional BSE-image, (**b**) elemental distribution maps and (**c**) quantitative line scans of Fe, Cr, and Al, as well as (**d**) XRD-pattern of Cr/Al-diffusion coated P91.

3.2. Exposure in Molten Solar Salt

The uncoated P91 specimen showed a markedly high specific net mass gain of 36.3 mg/cm^2 after 1000 h of exposure at 600 °C in MSS (Figure 5). Similar very high mass gain values were reported in other studies in the literature [2,10,12] for P91 and other 9% Cr steels (such as P92) during isothermal exposure in molten nitrates at 600 °C. On the other hand, coated specimens exhibited a significantly lower mass gain. In particular, the Al- and Cr/Al-coated specimens both showed almost negligible mass gains, which can be interpreted as the formation of a highly protective oxide scale. However, it should be noted that spallation effects cannot be revealed solely based on net mass change kinetics. Hence, the mass change kinetics should be evaluated together with the findings of analytical characterization in order to verify the improvement of the scaling behavior via coatings with respect to the uncoated samples.

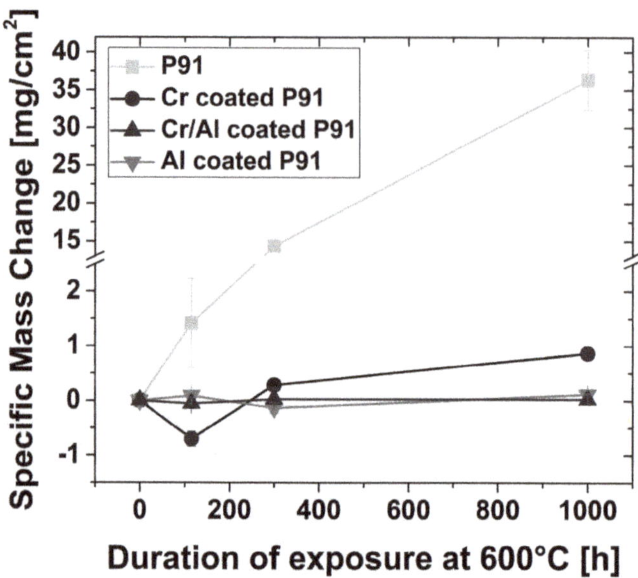

Figure 5. Specific mass change during isothermal exposure in molten solar salt at 600 °C. Please note the *y*-axis break.

Post-exposure XRD analysis showed the formation of hematite, magnetite, and sodium ferrite ($NaFeO_2$) on the uncoated P91 specimen after 1000 h exposure (Figure 6). Reflexes of $NaNO_3$ and KNO_3 were also detected, in addition to the corrosion products mentioned. This can be explained by the presence of salt residues, which apparently could not be removed from the sample surface upon rinsing in warm water and might possibly have also interfered with the determined value of specific mass change (Figure 5) to some extent. Nevertheless, taking into account the remarkably high specific mass change of the uncoated P91, this interference is considered to be very low. On the other hand, Cr_2O_3 was detected on the chromized P91 sample as well as hematite, magnetite, and sodium ferrite. XRD-patterns of Al- and Cr/Al-coated specimens exhibited the formation of corundum, hematite, and $FeAl_2O_4$ spinel as the oxide phases alongside the Fe-Al (as well as Cr_5Al_8 for the Cr/Al-coated sample) intermetallic phases, which were also present in the as-deposited coatings (see Figures 3 and 4). Within the diffraction pattern of Al- and Cr/Al-coated samples, the reflex at the two theta value of nearly 33.5° could not be identified. Due to its close location to the reflexes of Fe_2O_3 ($2\theta = 33.2°$) and Cr_2O_3 ($2\theta = 33.6°$), this reflex could originate from a distorted corundum structure. Owing to its high intensity, this reflex can be associated with the alumina inclusions at the surface of the coating, resulting from the in-pack cementation process [30,46]. Compressive residual stresses lead to lattice distortion and could explain the shift of the reflex compared to the stress-free state [47,48].

3.2.1. Scaling Behavior of Uncoated P91

An approximately 200 µm thick and porous oxide scale was formed on the uncoated P91 sample after 1000 h exposure in MSS at 600 °C, as shown in Figure 7. The oxide scale had a striated morphology and consisted of hematite, magnetite, and sodium ferrite, as identified by the post exposure XRD analysis (Figure 6). Interestingly, Cr was not present within the oxide scale, and underneath the oxide scale an approximately 40 µm thick Cr-depletion zone with a mean Cr-concentration of only 6 at.% was formed. This enhanced Cr-depletion from the alloy is explained by the transformation of Cr-rich oxide scales to chromate species and their dissolution into the salt melt [11,16]. With increasing exposure temperature, the decomposition of nitrate salts is accelerated, as highlighted in [1,8,14]. This leads to a higher equilibrium content of oxidizing agents in the salt melt, according to Reactions 1 and 2 and thereby enhanced Cr-depletion and thus an earlier onset of breakaway for P91 compared to lower exposure temperatures [26]. This is in good agreement with the study by Bradshaw and Carling, who reported an increasing extent of dissolved Cr cations in the salt melt with increasing temperature [8].

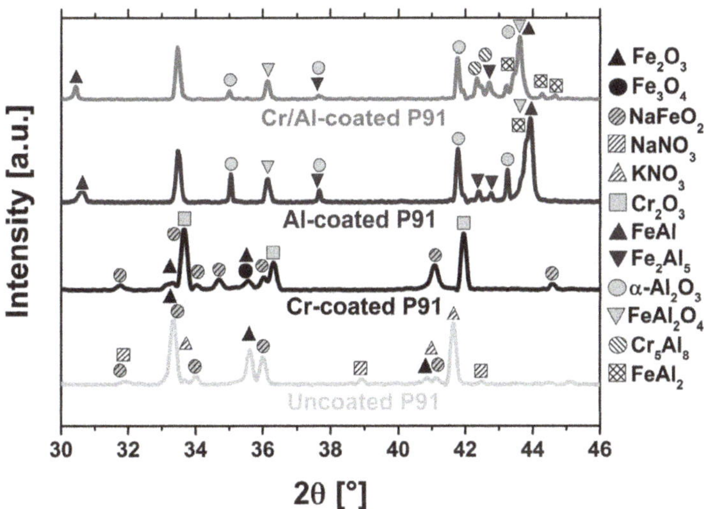

Figure 6. XRD patterns of uncoated, chromized, Al- and Cr/Al-coated P91 specimens after 1000 h exposure in molten solar salt at 600 °C.

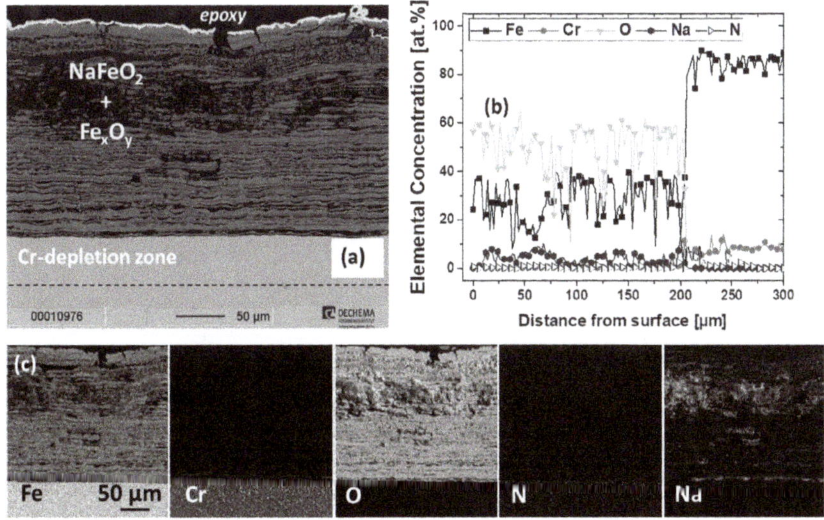

Figure 7. (**a**) Cross-sectional BSE-image, (**b**) quantitative line scans of Fe, Cr, O, Na, and N, and (**c**) elemental distribution maps of uncoated P91 after 1000 h isothermal exposure in molten solar salt at 600 °C.

Instead of local enrichment at the scale/melt interface, Na was present within the whole oxide scale (see Na map in Figure 7c) up to a maximum concentration close to 10 at.% (Figure 7b). As aforementioned, $NaFeO_2$ was detected via post-exposure XRD analysis (Figure 6). This is in good agreement with the study by Audigié et al. showing the formation of $NaFeO_2$ in the multi-layered oxide scale grown on P91 during exposure in molten nitrates at 580 °C [25]. The incorporation of Na into the oxide scales and the consequential formation of sodium ferrite is attributed to the transfer of Na cations from the salt melt to the scale surface and their inward penetration through oxide scale defects such as pores and cracks [18]. Moreover, N was enriched underneath the porous oxide scale (see Figure 7b and N map in Figure 7c), suggesting the nitridation of the sub-scale region, which was also observed in other studies investigating the molten nitrate corrosion of this alloy at 580 °C [6,25]. Nitrogen transport through the oxide scale to the alloy surface can be associated with oxide scale defects.

3.2.2. Scaling Behavior of Chromized P91

Figure 8 shows the cross-sectional BSE-image, elemental line scans, and distribution maps of the chromized P91 specimen after 1000 h exposure in MSS at 600 °C. In contrast to the 200 μm thick and highly non-protective oxide scale grown on the uncoated P91 specimen, the oxide scale grown on the chromized specimen was only 10 μm thick and consisted of an outer hematite and an inner Cr-rich oxide layer. The oxide scale formed at 600 °C in this study exhibited a very similar duplex morphology as the one grown on chromized P91 after exposure at 560 °C investigated in an earlier study [26]. Alongside the presence of hematite, Na was enriched within the outer layer of the oxide scale up to nearly 5 at.% (Figure 8b), which evidently led to the local formation of sodium ferrite, as identified by the XRD analysis (see Figure 6). In the study mentioned earlier, the Na-content in the oxide scale was significantly lower compared to this study and the formation of alkali ferrite species was excluded. Evidently, increased exposure temperature led to an enhanced Na inward flux from the melt to the scale, which can be associated with the increased basicity of the melt at higher temperatures, promoting the inward flux of Na cations [13]. At the scale/metal interface, a very thin and Cr-rich inner layer was formed. Post-exposure XRD analysis revealed the presence of Cr_2O_3 at the surface of the chromized specimen (Figure 6). The determined Cr-concentration of 17 at.% at the inner layer of the oxide scale (see Figure 8b) appeared to be low to stabilize Cr_2O_3. Evidently, the employed spatial resolution of 1 μm for EPMA line scans was not enough to clearly detect the Cr-concentration in the thin inner layer of the scale. Most likely, the EPMA signal was influenced by the surrounding regions, such as the gap at the interface between the oxide scale and the alloy surface, which was formed during metallographic preparation. Underneath the oxide scale, the as-deposited $Cr_{23}C_6$ layer (see Figure 2) was dissolved during exposure, but the Cr-diffusion zone with a mean Cr-concentration of 16 at.% was still visible after 1000 h exposure (Figure 8b). Notwithstanding, underneath the inner layer of the scale, Cr-depletion was observed to some extent, since the Cr-concentration dropped to 11.6 at.% at the surface of the coating (see Figure 8b). Nevertheless, the Cr-reservoir was evidently still adequate to counter-act the depletion of Cr to the salt, as was forecasted in an earlier study [26]. However, it should be mentioned that the oxide scale grown on the chromized P91 could not hinder the transfer of Na^+ cations from the salt melt to the scale. While improving the scaling behavior substantially (see Figure 5), the application of the Cr-diffusion coating did not alter the aforementioned corrosion mechanisms that led to a breakaway of uncoated P91 (see Figure 7). Evidently, the transfer of Na cations from the melt to the scale and that of Cr cations in the counter direction were restrained, but not inhibited. Furthermore, N was enriched within the diffusion zone to a mean concentration of 5 at.% (Figure 8b), suggesting nitridation of the alloy sub-surface. Its enrichment was more pronounced in the ferrite grains rather than in the needle shaped Cr- and Mo-rich carbide precipitates at the grain boundaries (see Figure 8c).

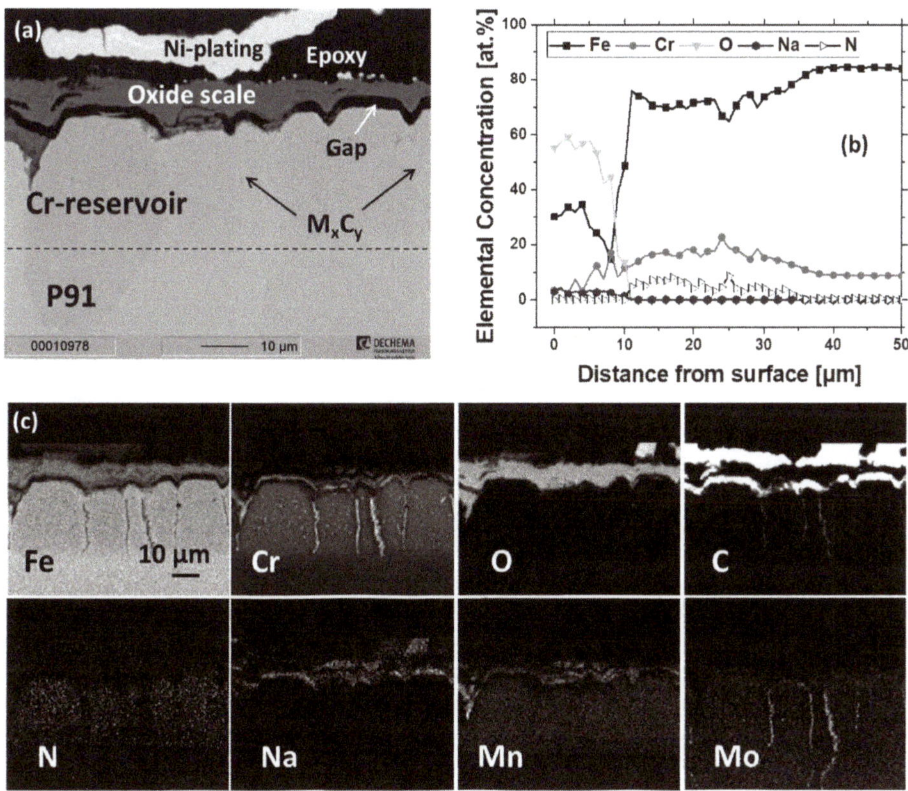

Figure 8. (**a**) Cross-sectional BSE-image, (**b**) quantitative line scans of Fe, Cr, O, Na, and N, and (**c**) elemental distribution maps of chromized P91 after 1000 h isothermal exposure in molten solar salt at 600 °C.

3.2.3. Scaling Behavior of Al-Coated P91

The Al-coating formed an even thinner oxide scale compared to the chromized P91 specimen and did not show substantial microstructural alteration as shown in Figure 9. The enhanced stability of aluminized P91 is in good correlation with other studies in literature [12,25,26], which reported improved scaling behavior for slurry aluminized P91 in molten nitrate salts. Post-exposure XRD analysis revealed that the oxide scale grown on the aluminized specimen after 1000 h exposure in MSS at 600 °C consisted of corundum, hematite and $FeAl_2O_4$ (see Figure 6). Na was detected in the oxide scale to an extent of ca. 4 at.% (Figure 9b) as well as in the cracks (see Na map in Figure 9c), which healed during exposure.

EPMA line scans revealed the sustainment of the as-deposited Al-concentration under the surface (see Figures 3c and 9b). Underneath the scale, the coating microstructure was almost identical to the as-deposited condition with an outer layer consisting of the Al-rich Fe-Al intermetallic phases ($FeAl_2$ and Fe_2Al_5) above an intermediate FeAl layer and an inner Al-diffusion zone (see Figures 3a and 9a). The cracks, which were present in the as-deposited condition, were healed during exposure, as can be best identified by the Al and O distribution maps in Figure 9c. Evidently, Kirkendall voids located at the interface between the FeAl layer and Al-diffusion zone in the as-deposited condition (see Figure 3a) did not grow during exposure (Figure 9a). Unlike its enrichment within the Cr-diffusion zone of the chromized specimen after 1000 h exposure (see Figure 8c), N was only detected in the nitride precipitates within the Al-diffusion zone (Figure 9c), which were also present in the as-deposited condition (see Figure 3b).

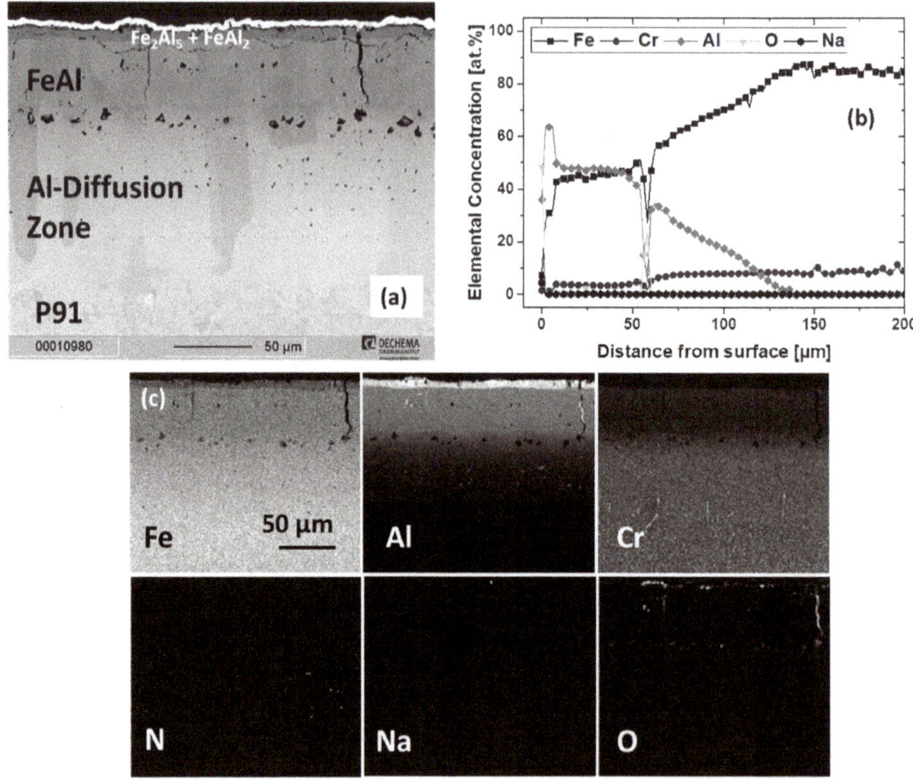

Figure 9. (**a**) Cross-sectional BSE-image, (**b**) quantitative line scans of Fe, Cr, Al, O and Na and (**c**) elemental distribution maps of aluminized P91 after 1000 h isothermal exposure in molten solar salt at 600 °C.

3.2.4. Scaling Behavior of Cr/Al-Coated P91

Cr/Al-coated P91 showed the lowest specific net mass gain (0.028 mg/cm^2) after 1000 h exposure in MSS at 600 °C. Correspondingly, a very thin and adherent oxide scale was formed on the Cr/Al-coating, as shown in Figure 10. Similar to the Al-coating (see Figure 9), the microstructure of the Cr/Al-coating was barely altered after 1000 h exposure. Again, analogous to the Al-coating, the oxide scale grown on the Cr/Al-coating consisted of corundum, hematite, and FeAl$_2$O$_4$, as determined by XRD analysis (Figure 6). Furthermore, Na was detected within the oxide scale up to a concentration of 8 at.% (Figure 10b). The elemental distribution maps showed more homogeneous and pronounced Na enrichment within the oxide scale (see Na and O maps in Figure 10c) compared to the aluminized P91 (Figure 9). Underneath the oxide scale, the Al-reservoir of the as-deposited coating was maintained (see Figures 4c and 10b) and hence, pristine coating phases, namely FeAl$_2$, Fe$_2$Al$_5$ and Cr$_5$Al$_8$ in the outer layer and FeAl in the inner layer, were still present (Figure 6). Cracks perpendicular to the surface, which were present in the as-deposited condition (see Figure 4a), were filled with a Na- and Al-rich oxide (see Na, Al and O maps in Figure 10c). Similar to the aluminized P91, instead of being enriched in the diffusion zone during exposure, N was only detected within the nitride precipitates in the Al-diffusion zone (Figure 10c), which were present in the as-deposited condition (see Figure 4b).

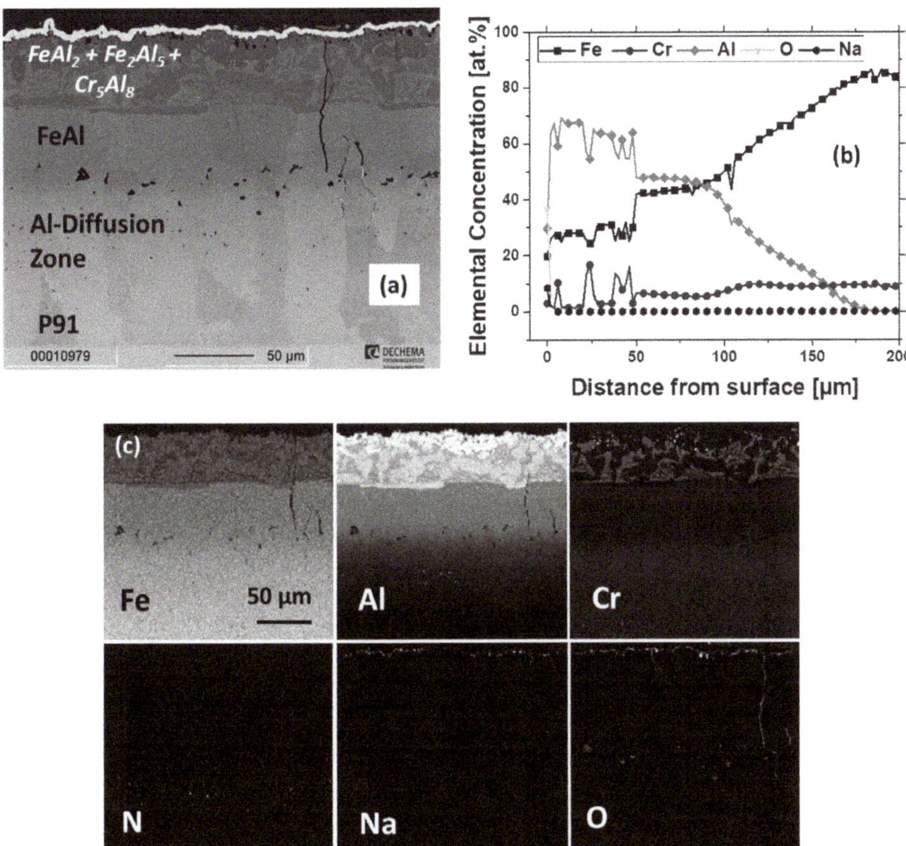

Figure 10. (**a**) Cross-sectional BSE-image, (**b**) quantitative line scans of Fe, Cr, Al, O, and Na, and (**c**) elemental distribution maps of Cr/Al-coated P91 after 1000 h isothermal exposure in molten solar salt at 600 °C.

3.2.5. Formation of Corundum on Al- and Cr/Al-Coated P91

The presence of corundum on both investigated aluminide coatings after 1000 h exposure in MSS at 600 °C was confirmed by post-exposure XRD analysis (Figure 6). However, the characterization of thin oxide scales grown on coated specimens during molten salt exposure via XRD can be difficult. This limitation is based on the low intensity of reflexes from very thin scales, as well as the overlapping of reflexes belonging to several oxide and intermetallic phases, as highlighted in [12,24]. Raman spectroscopy was therefore used to ascertain the composition of the oxide scales grown on Al- and Cr/Al-coated specimens. Figure 11a shows the surface morphology of the scales formed on aluminized P91. A large fraction of the surface, appearing dark in the secondary electron image (representatively marked as spot 1), is covered with a very thin oxide scale (see Figure 9). The corresponding Raman spectrum (Figure 11c) exhibited the characteristic fluorescence doublet of α-Al$_2$O$_3$ at a frequency around 1400 cm^{-1} [49]. Local protrusions, appearing bright in the secondary electron image (representatively marked as spot 2), were also observed on the surface. Such protrusions were clearly composed of α-Al$_2$O$_3$ (see Figure 11c). The higher relative intensity of the α-Al$_2$O$_3$ fluorescence doublet on spot 2 compared to spot 1 can be explained by a larger fraction of oxide phase in the acquired volume of spot 2, since the scale in spot 1 is very thin (see Figure 9a). Similarly, Figure 11b exhibits the surface morphology of the scales formed on Cr/Al-coated P91 after 1000 h exposure in MSS at 600 °C. The surface is covered with α-Al$_2$O$_3$ as confirmed with the α-Al$_2$O$_3$ fluorescence doublet (spot 1 in Figure 11d). Compared to the aluminized P91, fewer protrusions were observed on the surface of the Cr/Al-coated P91 (Figure 11b). These protrusions were also smaller, suggesting that the re-healing of a protective α-Al$_2$O$_3$ scale was much faster for the Cr/Al-coated than the Al-coated P91.

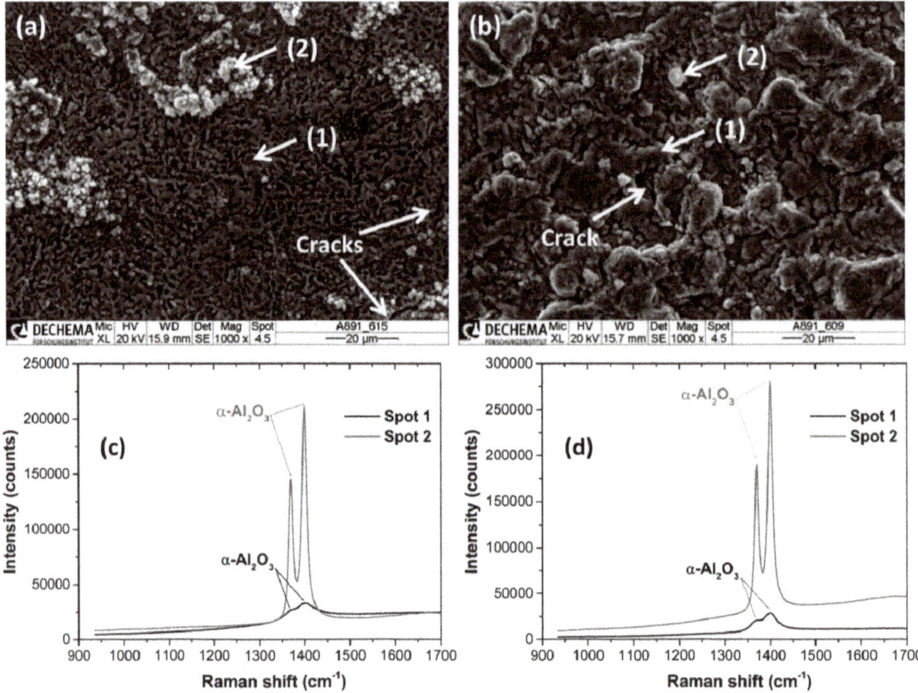

Figure 11. Oxidized surface morphologies and Raman spectra (fluorescence) from the marked spots of (**a,c**) aluminized and (**b,d**) Cr/Al-coated P91 after 1000 h isothermal exposure in molten solar salt at 600 °C.

To the authors' knowledge, the formation of corundum during molten nitrate exposure has not been reported in the literature. Usually, exposure of Fe-Al binary alloys at temperatures below 1000 °C in air is associated with the initial formation of metastable Al$_2$O$_3$ phases such as γ and θ, followed by their transformation to the slow growing corundum [50]. The temperature dependence of the time required for the transformation of metastable alumina phases to α-Al$_2$O$_3$ would lead to the requirement of significantly long exposure durations for the formation of corundum at temperatures such as 600 °C. It should be noted, that compared to binary alloys, diffusion aluminide coatings have a strictly different chemical composition, which is strongly affected by the chemical composition of the substrate. Their distinctive chemical composition including alloying elements might lead to different behavior with respect to binary alloys. As a matter of fact, Cr can promote the formation and sustainment of a protective corundum scale on ferritic as well as austenitic steels with a low Al-content via the so-called "third element effect", which is explained in detail elsewhere [51–54]. The sub-surface of the aluminide coating contained nearly 3 at.% Cr (see Figure 3c), whereas the Cr-content in the sub-surface of the Cr/Al-coating was substantially higher particularly due to the presence of Cr$_5$Al$_8$ precipitates (see Figure 4c). It could therefore be postulated, that the Cr-concentration at the sub-surface of both coatings might be sufficient to promote the formation and sustainment of a protective α-Al$_2$O$_3$

scale. In addition, despite the coating manufacturing process conducted in a reducing atmosphere (Ar + 5 vol.% H_2), one could expect the formation of a native Al_2O_3 scale on the surface of Al-rich intermetallic phases (i.e., $FeAl_2$, Fe_2Al_5 as well as Cr_5Al_8 for the Cr/Al-coating) considering the coating temperature around 1000 °C (see Table 2). The beneficial influence of the pre-oxidation of aluminide coatings upon aluminizing on the growth and sustainment of α-Al_2O_3 scales has been reported by Boulesteix et al. in steam exposure [49]. Furthermore, as aforementioned alumina particles can be incorporated into the coating due to the in-pack cementation process, as highlighted in [30]. The formation of a native Al_2O_3 scale along with the alumina inclusions from the pack cementation process might have acted as nucleation sites for the growth of corundum during exposure.

For both Al- and Cr/Al-coated P91, cracks can be observed through the oxide scales (Figure 11) and most likely correspond to areas adjacent to the vertical cracks observed in the cross-section images in Figures 9a and 10a. These cracks were not associated with the spallation of the oxide scales, suggesting that the α-Al_2O_3 scales were strongly adherent to the underlying intermetallic coatings. The microstructure of both Al- and Cr/Al-coatings was not degraded during exposure in molten nitrates (see Figures 3 and 4 for the as-deposited condition and Figures 9 and 10 after exposure). This suggests that a very low extent of outward Al diffusion was sufficient to form a thin and adherent α-Al_2O_3 scale. Alumina, as a well-known insulator, can hinder the charge transfer between the specimen and the electrolyte. The observed protrusions (Figure 11) might have originated due to the local defects in the oxide scale and local increase in the electrochemical activity.

Notwithstanding the formation of corundum, it should be noted that Na-containing compounds such as $NaAlO_2$ might have formed during the exposure, as highlighted in [16,24]. Tortorelli et al. conducted thermodynamic calculations and reported the stability of $NaAlO_2$ for a wide $p(O_2)$-$p(N_2)$ range at relevant temperatures for CSP operation [16]. These authors associated the high corrosion resistance of iron aluminides to the formation of a sodium aluminate scale and its low solubility in the salt melt. In this study, neither sodium aluminate nor any other Na-containing oxide phases were detected via post-exposure XRD analysis or Raman spectroscopy. This can be associated with the rinsing of the samples after exposure, galvanic Ni-plating, and subsequent metallographic preparation, which might have led to their dissolution, as highlighted in [16]. Hence the origin of Na enrichment in the oxide scale cannot definitely be linked to the formation of $NaAlO_2$ with the employed analytical methods in this study. It should also be noted, that according to the binary Na_2O-Al_2O_3 phase diagram [55], sodium has a solubility limit of nearly 3 at.% in corundum and increasing Na-concentration leads to the stabilization of β-Al_2O_3 ($NaAl_{11}O_{17}$) and $NaAlO_2$ phases.

The possible formation of sodium aluminate can be associated with the basic dissolution of Al_2O_3 in a melt with high O^{2-} activity such as molten nitrates, following the reactions (4) and (5):

$$Al_2O_3(s) + O^{2-} \rightleftharpoons 2AlO_2^{-} \tag{4}$$

$$Na^{+} + AlO_2^{-} \rightleftharpoons NaAlO_2(s) \tag{5}$$

The overall dissolution reaction can therefore be written as:

$$Al_2O_3(s) + Na_2O \rightleftharpoons 2NaAlO_2(s) \tag{6}$$

The same mechanisms were proposed by Audigié et al. to explain the formation of sodium ferrite from hematite [25]. Despite no direct observation of sodium aluminate on Al- and Cr/Al-coated samples, the formation of a thin α-Al_2O_3 scale and possibly $NaAlO_2$ on the surfaces were responsible for the superior corrosion resistance of the Al-rich coatings in MSS at 600 °C, as first proposed by Tortorelli et al. [16].

4. Conclusions

The scaling behavior of Cr-, Al- and Cr/Al-coatings deposited on P91 in MSS at 600 °C was studied with respect to uncoated P91. The investigated diffusion coatings significantly increased the corrosion resistance of P91 in the above-mentioned CSP relevant exposure conditions. Hence, diffusion coatings can enable the employment of ferritic-martensitic steels as structural materials in CSP systems with TES, which can increase the cost-efficiency due to their lower unit price compared to austenitic steels and Ni-based alloys. The additional costs associated with the coating processes e.g., pack cementation are economically viable considering the notable increase in lifetime of ferritic-martensitic steels in MSS via coatings.

The oxide scale grown on uncoated P91 exposed in MSS at 600 °C for 1000 h was nearly 200 μm thick and exhibited a porous and striated morphology which consisted of hematite, magnetite, and sodium ferrite. Na was detected over the whole thickness of the oxide scale to a concentration close to 10 at.%. Notably, Cr was completely depleted from the oxide scale, suggesting the formation of chromate species and their dissolution in the salt melt. In contrast, the oxide scale grown on the chromized specimen was only 10 μm thick and possessed a Cr-rich inner layer alongside the outer hematite layer. Compared to the uncoated P91, sodium ferrite was detected to a lesser extent at the surface of the chromized specimen. The incorporation of Na into the oxide scale suggested that similar corrosion mechanisms prevailed for both uncoated and chromized P91 specimens. However, despite nitridation of the sub-surface to some extent, the Cr-reservoir was maintained and thus breakaway of P91 in MSS was shifted to prolonged exposure durations via the Cr-diffusion coating.

Al- and Cr/Al-coatings exhibited an almost unaltered microstructure due to the very low extent of interdiffusion at the exposure temperature, as well as an evidently very low extent of outward Al diffusion being sufficient to form a highly protective oxide scale. The scale morphology was more homogeneous in the case of the Cr/Al-coating, which was associated with the presence of Cr_5Al_8 precipitates increasing the Al-activity at the sub-surface enabling faster healing of the oxide scale as well as promoting corundum formation via the third element effect. Raman spectroscopy verified the formation of corundum on both coatings at the relatively low exposure temperature. The highly protective behavior of aluminide coatings was associated with corundum being an electrical insulator and thus hindering charge transfer between the electrolyte and sample surface. However, it should be noted that the formation of $NaAlO_2$ alongside corundum cannot be excluded, since the metallographic preparation of the samples might have led to its dissolution. Nevertheless, due to its low solubility in the salt melt, the possible formation of sodium aluminate on the aluminide coated specimens should not deteriorate the protective behavior, which is expected to be sustained for prolonged exposure durations.

5. Outlook

The formation of hexavalent Cr cations during exposure in molten nitrate salts should be considered as an important environmental issue due to chromates being highly toxic, especially considering the amount of salt utilized in CSP plants. Therefore, the determination of metal cations including Cr^{6+} and the investigation of nitrite content in the solar salt during exposure of coated as well as uncoated alloys is to be studied in future. Moreover, the corrosion behavior of alumina forming stainless steels at the studied temperature range is considered to be interesting for CSP applications, similar to the investigation at a lower temperature of 390 °C in [56]. Furthermore, the influence of diffusion coatings on the mechanical behavior of P91 should be of future interest to elucidate its employment in CSP plants. The formation of a large ferritic zone due to the inward diffusion of ferrite formers e.g., Al, refractory carbide precipitates in the as-chromized microstructure as well as cracks in the as-deposited condition of aluminide coatings are expected to influence the mechanical behavior of P91, which will be investigated in a future study.

Author Contributions: Conceptualization, C.O., T.M.M. and M.C.G.; Methodology, C.O., T.M.M., C.D. and B.G.; Investigation, C.O., T.M.M., C.D. and B.G.; Data Curation, C.O., T.M.M., C.D. and B.G.; Writing–Original Draft Preparation, C.O., T.M.M. and B.G.; Writing–Review and Editing, C.O., T.M.M., B.G. and M.C.G.

Funding: This research has received funding from the European Union's Horizon 2020 research and innovation program under grant agreement No. 686008 (RAISELIFE).

Acknowledgments: The authors would like to acknowledge Gerald Schmidt for the EPMA measurements, Nadine Thuma for the metallographic preparation, and Rick N. Durham for proofreading. Furthermore, the authors would like to thank all project partners from the "RAISELIFE" project for helpful scientific discussions.

Conflicts of Interest: The authors declare no conflict of interest.

References

1. Bonk, A.; Sau, S.; Uranga, N.; Hernaiz, M.; Bauer, T. Advanced heat transfer fluids for direct molten salt line-focusing CSP plants. *Prog. Energy Combust. Sci.* **2018**, *67*, 69–87. [CrossRef]
2. Soleimani-Dorcheh, A.; Durham, R.N.; Galetz, M.C. Corrosion behavior of stainless and low-chromium steels and IN625 in molten nitrate salts at 600 C. *Sol. Energy Mater. Sol. Cells* **2016**, *144*, 109–116. [CrossRef]
3. Barlev, D.; Vidu, R.; Stroeve, P. Innovation in concentrated solar power. *Sol. Energy Mater. Sol. Cells* **2011**, *95*, 2703–2725. [CrossRef]
4. Vignarooban, K.; Xu, X.; Arvay, A.; Hsu, K.; Kannan, A.M. Heat transfer fluids for concentrating solar power systems—A review. *Appl. Energy* **2015**, *146*, 383–396. [CrossRef]
5. Boretti, A.; Casteletto, S.; Al-Zubaidy, S. Concentrating solar power tower technology: Present status and outlook. *Nonlinear Eng.* **2019**, *8*, 10–31. [CrossRef]
6. Agüero, A.; Audigié, P.; Rodríguez, S.; Encinas-Sánchez, V.; de Miguel, M.T.; Pérez, F.J. Protective coatings for high temperature molten salt heat storage systems in solar concentration power plants. In *AIP Conference Proceedings*; AIP Publishing: Melville, NY, USA, 2018; Volume 2033.
7. Bonk, A.; Braun, M.; Hanke, A.; Forstner, J.; Rückle, D.; Kaesche, S.; Sötz, V.A.; Bauer, T. Influence of different atmospheres on molten salt chemistry and its effect on steel corrosion. In *AIP Conference Proceedings*; AIP Publishing: Melville, NY, USA, 2018; Volume 2033.
8. Bradshaw, R.W.; Carling, R.W. A review of the chemical and physical properties of molten alkali nitrate salts and their effect on materials used for solar central receivers. *Proc. Electrochem. Soc.* **1987**, *1987*, 959–969. [CrossRef]
9. Kruizenga, A.; Gill, D. Corrosion of iron stainless steels in molten nitrate salt. *Energy Procedia* **2014**, *49*, 878–887. [CrossRef]
10. Spiegel, M.; Mentz, J. High temperature corrosion beneath nitrate melts. *Mater. Corros.* **2014**, *65*, 276–281. [CrossRef]
11. Summers, K.L.; Chidambaram, D. Corrosion behavior of structural materials for potential use in nitrate salts based solar thermal power plants. *J. Electrochem. Soc.* **2017**, *164*, 5357–5363. [CrossRef]
12. Soleimani-Dorcheh, A.; Galetz, M.C. Slurry aluminizing: A solution for molten nitrate salt corrosion in concentrated solar power plants. *Sol. Energy Mater. Sol. Cells* **2016**, *146*, 8–15. [CrossRef]
13. Fernandez, A.G.; Lasanta, M.I.; Perez, F.J. Molten salt corrosion of stainless steels and low-Cr steel in CSP plants. *Oxid. Met.* **2012**, *78*, 329–348. [CrossRef]
14. McConohy, G.; Kruizenga, A. Molten nitrate salts at 600 and 680 C: Thermophysical property changes and corrosion of high-temperature nickel alloys. *Sol. Energy* **2014**, *103*, 242–252. [CrossRef]
15. Rapp, R.A.; Goto, K.S. The hot corrosion of metals by molten salts. In Proceedings of the Second International Symposium on Molten Salts; Selman, R., Braunstein, J., Eds.; Electrochemical Society: Pennington, NJ, USA, 1981; pp. 159–177.
16. Tortorelli, P.F.; Bishop, P.S.; DiStefano, J.R. *Selection of Corrosion-Resistant Materials for Use in Molten Nitrate Salts ORNL/TM-11162*; Oak Ridge National Lab.: Oak Ridge, TN, USA, 1989.
17. Bradshaw, R.W.; Clift, W.M. *Effect of Chloride Content of Molten Nitrate Salt on Corrosion of A516 Carbon Steel*; Sandia Report, SAND2010-7594; Sandia National Laboratories: Albuquerque, NM, USA; Livermore, CA, USA, 2010.
18. Dorcheh, A.S.; Durham, R.N.; Galetz, M.C. High temperature corrosion in molten solar salt: The role of chloride impurities. *Mater. Corros.* **2017**, *68*, 943–951. [CrossRef]

19. Encinas-Sánchez, V.; de Miguel, M.T.; Lasanta, M.I.; García-Martín, G.; Pérez, F.J. Electrochemical impedance spectroscopy (EIS): An efficient technique for monitoring corrosion processes in molten salt environments in CSP applications. *Sol. Energy Mater. Sol. Cells* **2019**, *191*, 157–163. [CrossRef]

20. Bradshaw, R.W.; Goods, S.H. *Corrosion Resistance of Stainless Steels During Thermal Cycling in Alkali Nitrate Molten Salts*; Sandia Report, SAND2001-8518; Sandia National Laboratories: Albuquerque, NM, USA; Livermore, CA, USA, 2001.

21. Goods, S.H.; Bradshaw, R.W. Corrosion of stainless steels and carbon steel by molten mixtures of commercial nitrate salts. *J. Mater. Eng. Perform.* **2004**, *13*, 78–87. [CrossRef]

22. Slusser, J.W.; Titcomb, J.B.; Heffelfinger, M.T.; Dunbobbin, B.R. Corrosion in molten nitrate-nitrite salts. *JOM* **1985**, *37*, 24–27. [CrossRef]

23. Fähsing, D. Ferritisch Martensitische Stähle: Verbesserung der Oxidationsbeständigkeit in Wasserdampfhaltigen Hochtemperaturatmosphären. Ph.D. Thesis, RWTH Aachen, Shaker Verlag, Herzogenrath, Germany, 2015.

24. Audigié, P.; Bizien, N.; Baráibar, I.; Rodriguez, S.; Pastor, A.; Hernández, M.; Agüero, A. Aluminide slurry coatings for protection of ferritic steel in molten nitrate corrosion for concentrated solar power technology. In *AIP Conference Proceedings*; AIP Publishing: Melville, NY, USA, 2017; Volume 1850, p. 070002.

25. Audigié, P.; Encinas-Sánchez, V.; Juez-Lorenzo, M.; Rodríguez, S.; Gutiérrez, M.; Pérez, F.J.; Agüero, A. High temperature molten salt corrosion behavior of aluminide and nickel-aluminide coatings for heat storage in concentrated solar power plants. *Surf. Coat. Technol.* **2018**, *349*, 1148–1157. [CrossRef]

26. Fähsing, D.; Oskay, C.; Meißner, T.M.; Galetz, M.C. Corrosion testing of diffusion-coated steel in molten salt for concentrated solar power tower systems. *Surf. Coat. Technol.* **2018**, *354*, 46–55. [CrossRef]

27. *Material Data Sheet P91/T91*; Thyssen Krupp Materials International: Essen, Germany, 2011.

28. Goward, G.W. Protective Coatings for High Temperature Alloys: State of Technology. In Proceedings of the Symposium on Properties of High-Temperature Alloys with Emphasis on Environmental Effects, New York, NY, USA, 17 October 1976.

29. Duret, C.; Pichoir, R. Protective Coatings for High Temperature Materials: Chemical Vapour Deposition and Pack Cementation Processes. In *Coatings for High Temperature Applications*; Lang, E., Ed.; Applied Science Publishers: London, UK, 1983; pp. 33–78.

30. Mevrel, R.; Duret, C.; Pichoir, R. Pack cementation processes. *Mater. Sci. Technol.* **1986**, *2*, 201–206. [CrossRef]

31. Rohr, V.; Donchev, A.; Schütze, M.; Milewska, A.; Pérez, F.J. Diffusion coatings for high temperature corrosion protection of 9–12% Cr steels. *Corros. Eng. Sci. Technol.* **2005**, *40*, 226–232. [CrossRef]

32. *ISO 17245:2015 Corrosion of Metals and Alloys—Test Method for High Temperature Corrosion Testing of Metallic Materials by Immersing in Molten Salt or Other Liquids under Static Conditions*; ISO: Geneva, Switzerland, 2015.

33. Schmidt, D.; Galetz, M.C.; Schütze, M. Ferritic–martensitic steels: Improvement of the oxidation behavior in steam environments via diffusion coatings. *Surf. Coat. Technol.* **2013**, *237*, 23–29. [CrossRef]

34. Bianco, R.; Rapp, R.A. Pack cementation diffusion coatings. In *Metallurgical and Ceramic Protective Coatings*; Stern, K.H., Ed.; Chapman & Hall: London, UK, 1996; pp. 236–260.

35. Meier, G.H.; Cheng, C.; Perkins, R.A.; Bakker, W. Diffusion chromizing of ferrous alloys. *Surf. Coat. Technol.* **1989**, *39*, 53–64. [CrossRef]

36. Agüero, A.; Muelas, R.; Pastor, A.; Osgerby, S. Long exposure steam oxidation testing and mechanical properties of slurry aluminide coatings for steam turbine components. *Surf. Coat. Technol.* **2005**, *200*, 1219–1224. [CrossRef]

37. Bates, B.L.; Zhang, Y.; Dryepondt, S.; Pint, B.A. Creep behavior of pack cementation aluminide coatings on grade 91 ferritic–martensitic alloy. *Surf. Coat. Technol.* **2014**, *240*, 32–39. [CrossRef]

38. Rohr, V.; Schütze, M.; Fortuna, E.; Tsipas, D.N.; Milewska, A.; Pérez, F.J. Development of novel diffusion coatings for 9–12% Cr ferritic-martensitic steels. *Mater. Corros.* **2005**, *56*, 874–881. [CrossRef]

39. Zhang, Y.; Pint, B.A.; Cooley, K.M.; Haynes, J.A. Formation of aluminide coatings on Fe-based alloys by chemical vapor deposition. *Surf. Coat. Technol.* **2008**, *202*, 3839–3849. [CrossRef]

40. Naji, A.; Galetz, M.C.; Schütze, M. Improvements in the thermodynamic and kinetic considerations on the coating design for diffusion coatings formed via pack cementation. *Mater. Corros.* **2015**, *66*, 863–868. [CrossRef]

41. Touloukian, Y.S.; Kirby, R.K.; Taylor, R.E.; Desai, P.D. *Thermal Expansion of Metallic Elements and Alloy*; Plenum Publishing Corporation: New York, NY, USA, 1970.

42. Fähsing, D.; Rudolphi, M.; Konrad, L.; Galetz, M.C. Fireside Corrosion of Chromium-and Aluminum-Coated Ferritic–Martensitic Steels. *Oxid. Met.* **2017**, *88*, 155–164. [CrossRef]

43. Zhang, Y.; Pint, B.A.; Cooley, K.M.; Haynes, J.A. Effect of nitrogen on the formation and oxidation behavior of iron aluminide coatings. *Surf. Coat. Technol.* **2005**, *200*, 1231–1235. [CrossRef]

44. Palm, M. The Al–Cr–Fe system–Phases and phase equilibria in the Al-rich corner. *J. Alloy Compd.* **1997**, *252*, 192–200. [CrossRef]

45. Pavlyuchkov, D.; Przepiórzynski, B.; Kowalski, W.; Velikanova, T.Y.; Grushko, B. Al–Cr–Fe phase diagram. Isothermal Sections in the region above 50 at% Al. *Calphad* **2014**, *45*, 194–203. [CrossRef]

46. Das, D.K.; Singh, V.; Joshi, S.V. Evolution of aluminide coating microstructure on nickel-base cast superalloy CM-247 in a single-step high-activity aluminizing process. *Metall. Mater. Trans. A* **1998**, *29*, 2173–2188. [CrossRef]

47. Fitzpatrick, M.E.; Fry, A.T.; Holdway, P.; Kandil, F.A.; Shackleton, J.; Suominen, L. *Determination of Residual Stresses by X-ray Diffraction Measurement Good Practice Guide 52*; National Physical Laboratory: Teddington, Middlesex, UK, 2005.

48. Kamminga, J.D.; de Keijser, T.H.; Mittemeijer, E.J.; Delhez, R. New methods for diffraction stress measurement: A critical evaluation of new and existing methods. *J. Appl. Cryst.* **2000**, *33*, 1059–1066. [CrossRef]

49. Boulesteix, C.; Kolarik, V.; Pedraza, F. Steam oxidation of aluminide coatings under high pressure and for long exposures. *Corros. Sci.* **2018**, *144*, 328–338. [CrossRef]

50. Grabke, H.J. Oxidation of NiAl and FeAl. *Intermetallics* **1999**, *7*, 1153–1158. [CrossRef]

51. Stott, F.H.; Wood, G.C.; Stringer, J. The influence of alloying elements on the development and maintenance of protective scales. *Oxid. Met.* **1995**, *44*, 113–145. [CrossRef]

52. Brady, M.P.; Yamamoto, Y.; Santella, M.L.; Walker, L.R. Composition, microstructure, and water vapor effects on internal/external oxidation of alumina-forming austenitic stainless steels. *Oxid. Met.* **2009**, *72*, 311–333. [CrossRef]

53. Heinonen, M.H.; Kokko, K.; Punkkinen, M.P.J.; Nurmi, E.; Kollár, J.; Vitos, L. Initial oxidation of Fe–Al and Fe–Cr–Al alloys: Cr as an alumina booster. *Oxid. Met.* **2011**, *76*, 331–346. [CrossRef]

54. Renusch, D.; Grimsditch, M.; Koshelev, I.; Veal, B.W. Strain Determination in Thermally-Grown Alumina Scales Using Fluorescence Spectroscopy. *Oxid. Met.* **1997**, *48*, 471–795. [CrossRef]

55. Lambotte, G.; Chartrand, P. Thermodynamic modeling of the $(Al_2O_3 + Na_2O)$, $(Al_2O_3 + Na_2O + SiO_2)$, and $(Al_2O_3 + Na_2O + AlF_3 + NaF)$ systems. *J. Chem. Thermodyn.* **2013**, *57*, 306–334. [CrossRef]

56. Fernandez, A.G.; Rey, A.; Lasanta, I.; Mato, S.; Brady, M.P.; Perez, F.J. Corrosion of alumina-forming austenitic steel in molten nitrate salts by gravimetric analysis and impedance spectroscopy. *Mater. Corros.* **2014**, *65*, 267–275. [CrossRef]

 *coatings*

Article

The Synthesis of a Superhydrophobic and Thermal Stable Silica Coating via Sol-Gel Process

Karmele Vidal, Estíbaliz Gómez, Amaia Martínez Goitandia, Adrián Angulo-Ibáñez and Estíbaliz Aranzabe *

IK4-Tekniker, Surface Chemistry and Nanotechnology Unit, Calle Iñaki Goenaga 5, 20600 Gipuzkoa, Spain; karmele.vidal@tekniker.es (K.V.); estibaliz.gomez@tekniker.es (E.G.); amaia.martinez@tekniker.es (A.M.G.); adrian.angulo@tekniker.es (A.A.-I.)

* Correspondence: estibaliz.aranzabe@tekniker.es; Tel.: +34-943206744

Received: 27 June 2019; Accepted: 26 September 2019; Published: 28 September 2019

Abstract: A super-hydrophobic surface at a high temperature (400 °C) using the sol-gel method with tetraethoxysilane (TEOS) and methyltriethoxysilane (MTES) as precursors has been obtained. The effects of the coatings' ages, deposited times and thicknesses on the hydrophobicity of the silica coatings have been analysed. The morphology, chemical composition, thermal degradation and hydrophobicity of the resulting surfaces have been studied by scanning electron microscopy (SEM), Fourier transfer infrared spectrometer (FT-IR), Thermogravimetry (TGA) and water contact angle (WCA) measurement. The results show that an average water contact angle of 149° after been cured at 400 °C for a coating aged for 5 days, and four deposition cycles using a dipping rate of 1000 mm/min was achieved.

Keywords: superhydrophobic surface; sol-gel; silica coating; temperature stability

1. Introduction

The phenomenon of superhydrophobicity, also known as the "Lotus Effect" in a material, occurs when water droplets roll easily on its surface with a slip angle of less than 10° giving rise to a water contact angle (WCA) higher than 150°. This phenomenon occurs due to the reduction of the water contact area, which leads to a reduction of adhesion [1,2].

Several works have been reporting about the preparation of superhydrophobic surfaces and their potential applications (anti-corrosion, self-cleaning, anticontamination, non-stick wares, etc.) [3,4].

Artificial superhydrophobic surfaces can be achieved by: (i) creating a rough surface on a hydrophobic material or (ii) chemically modifying a rough surface with a hydrophobic coating [5]. For this, different methods have been used: lithographing, templating method, phase separation, plasma treatment, sputter coating, solution-immersion process, chemical vapor deposition, physical etching and sol-gel reaction, hydrothermal synthesis, and others [6,7].

Silica coatings prepared with TEOS are inherently hydrophilic due to the presence of surface active hydroxyl (–OH) end groups. However, the hydrophobic behaviour can be improved by replacing the –OH group with the hydrophobic organic functional group, O–Si–$(CH_3)_3$. For this purpose, the most employed precursors in the sol-gel synthetic route are methyltrimethoxysilane (MTMS), methyltriethoxysilane (MTES) and vinyltriethoxysilane (VTES) [8–19]. Guilong et al., fabricated a superhydrophobic SiO_2 film with TEOS and MTES as precursors by the sol-gel method, and then, the SiO_2 particles obtained were modified by dodecyltrimethoxysilane (DTMS) [20].

For these superhydrophobic coatings to be implemented to industry, they must be easily fabricated and durable during services. However, many superhydrophobic coating materials deteriorate at relatively low temperatures, 300–350 °C, due to raw materials' low thermal stabilities, giving rise to a surface morphology or chemistry damage of the coatings surface [21–23].

The sol-gel synthetic route is appropriate to obtain superhydrophobic coatings with good properties and good thermal resistance [24–26]. Traditionally, most of these thin films have been prepared by sol-gel method because of it is a simple process and allows one to change the parameters to obtain the desired products [5].

On the other hand, generally, the studies of superhydrophobic surfaces are focused on rigid solid substrates, such as silicon wafers, glass slides and metal surfaces. These materials must be selected considering both the treatment processes and their final application [3,27].

In this work, a porous superhydrophobic and-stable-at-high-temperatures (~400 °C) silica film was synthesized via sol-gel method. To enhance the water repellence of the silica films, methyltriethoxysilane (MTES) was used as a hydrophobic reagent in the sol-gel process. The paper describes the influence of the aging period, dipping times and the thickness of coatings on the superhydrophobic, morphological, chemical and thermal stability properties of the silica films obtained using glass substrates. This investigation provides the optimal parameters to achieve, via sol-gel, superhydrophobic coatings suitable for high-temperature applications.

2. Materials and Methods

2.1. Materials

Tetraethyl orthosilicate (TEOS, 98%), methyltriethoxysilane (MTES, 98%) and ammonium hydroxide solution (ACS reagent, 28% to 30% NH_3 basis) were purchased from Sigma Aldrich (Saint Louis, MO, USA). Absolute ethanol (EtOH, 99.5%) and distilled water (H_2O) from Scharlab (Scharlab, S.L., Barcelona, Spain) were used in this work.

2.2. Preparation of Superhydrophobic Coating

Silica particles were prepared through a process based on the Stöber method [28], using tetraethoxysilane (TEOS) as a starting material. TEOS was dissolved in ethanol. Separately, ammonia, water and ethanol were mixed. Then, two solutions were mixed, and heated at 60 °C under stirring for 90 min. Then, a solution of MTES and EtOH was added to the sol and the mixture was continuously stirred for 19 h at 60 °C with a condensation reflux device. The molar ratio of MTES:TEOS:H_2O:EtOH:NH_4OH was 0.16:0.24:4:14:1. The prepared white silica sol was kept for 2, 5 and 9 days under room temperature.

Glass substrates were vertically dipped using an immersion-extraction velocity of 200 mm/min into test beakers containing the sol solution for 0 and 5 min, repeating the procedure 1 and 4 times with a thermal treatment at 150 °C for 30 min after each process. Adding to this, the dipping velocity was also varied for the sol aged for 5 days; velocities of 500 and 1000 mm/min were used. Finally, the substrates obtained were annealed at 400 °C for 2 h. The thermal stability was confirmed by putting the SiO_2 coating samples in an oven at different temperatures (400, 450, 500 and 550 °C) for 2 h, and then, testing the water contact angles for each coating. In addition, the dipping rate was analysed in order to study the effect of the thickness on the hydrophobic properties.

2.3. Characterization Techniques

Fourier transform infrared spectroscopy (FTIR, FT/IR-4700LE, JASCO International Co., Ltd., Tokyo, Japan) was used in order to obtain information about the carbon–hydrogen, silica–oxygen and functional group bonds. FTIR was operated in the 4000–400 cm^{-1} range and the average of three scans for each sample was taken for the peak identification.

A goniometer (SURFTENS Universal, OEG GmbH, Frankfurt, Germany) was used to observe the hydrophobicity of the silica coatings by measuring the contact angle (θ) and sliding angle (α) of a water droplet of 5 µL placed on the surface of the coated glass substrates. The sliding angle was measured by tilting the stage table of the goniometer. The average value of five measurements, made at different positions of the coating surface was adopted as the value of WCA.

Morphologies of the samples were observed using a scanning electron microscope Ultra Gemini-II de Carl Zeiss SMT (Thornwood, NY, USA). The thickness of the coatings was determined with a profilometer (Dektak 8, Karlsruhe, Germany).

Thermal degradation of coating material was carried out with thermal gravimetric analysis (TGA) in a TGA/DSC1 equipment from Mettler Toledo (GIEβen, Hesse, Germany) from room temperature up to 1000 °C at 10 °C/min.

3. Results and Discussion

3.1. Fourier Transform Infrared Spectroscopy

The effect of the addition of MTES and the evolution of the condensation of the sol were studied by FTIR. The Figure 1 shows the FTIR spectra of sol before and after the addition of MTES (incorporation of MTES to the reaction for 2, 5 and 9 days). A broad intense band is detected between 3000 and 3800 cm^{-1}, with a peak at 3323 cm^{-1}, due to O–H vibrations and a weak deformation vibrations at 1649 cm^{-1}. The 3323 cm^{-1} and Si–OH (953 cm^{-1}) bands decrease gradually during condensation reaction. In addition, two sharp bands at 2971 cm^{-1} and 1272 cm^{-1} corresponding to the absorption vibration of C–H in methyl and Si–CH$_3$ groups, respectively, indicate the substitution of hydroxil groups for methyl groups when MTES is added to the reaction. C–H bondsweare detected at 2928 cm^{-1} in a symmetric stretching mode and at 2971 cm^{-1} in an asymmetric stretching mode. The presence of the methyl groups is beneficial for the hydrophobic properties.

The Si–O–Si symmetric stretching vibration mode observed at 1048 cm^{-1} and the asymmetric mode at 1084 cm^{-1} indicate the hybrid nature of the material, which is in good agreement with other works [20,29]. When MTES is incorporated to the reaction, the bands increase due to the increase of Si–OH groups, and therefore, more Si–O–Si groups are formed during the condensation steps.

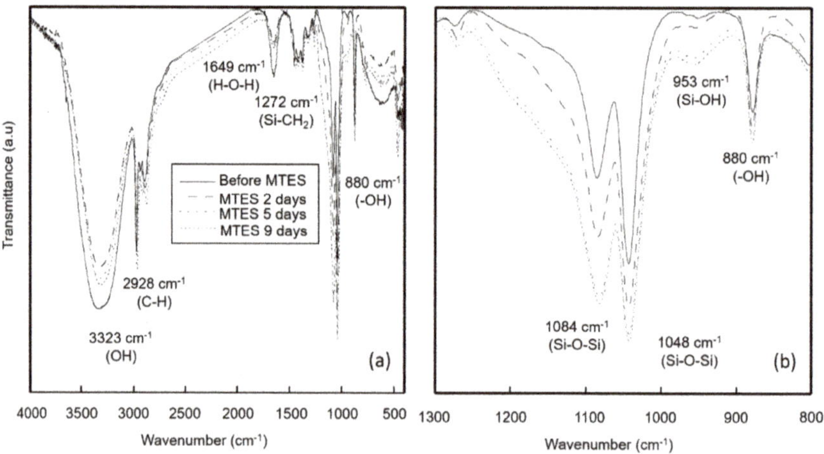

Figure 1. FTIR spectra of the sol before and after (2, 5 and 9 days) the addition of MTES.

A band of medium intensity at 953 cm^{-1} corresponds to Si–OH groups. However, this band is not univocally atributed to Si–OH modes. It can be overlapped with the Si–O of unreacted alcoxy groups from TEOS and MTES [30]. A weak Si–OH peak before MTES addition is observable, due to a complete hydrolisis and condensation of TEOS. After two days of MTES incorporation, the peak is sharper as a result of the presence of hydroxyl groups formed during the hydrolysis step, decreasing with time because of the condensation reaction and the transformation of these groups into Si–O–Si species.

The FTIR spectrum of the optimized hydrophobic SiO$_2$ coatings prepared by TEOS and MTES as precursors and aged for 5 days and cured at 400 °C is shown in Figure 2.

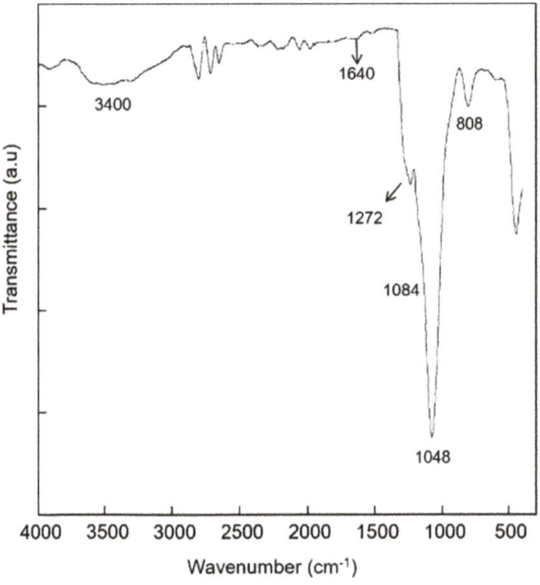

Figure 2. FTIR spectra of the SiO$_2$ particles for coating aged for 5 days.

As in the FTIR of the sol, the broad absorption bands at around 1649 and 3323 cm^{-1} indicate the presence of –OH groups. The stretching vibrations of the Si–O–Si bonds are represented by the 1048 cm^{-1} and 1084 cm^{-1} absorption bands. The presence of this peak confirms the formation of a network structure inside the coating. The peak at 1272 cm^{-1} indicates Si–CH$_3$ bonds due to the presence of MTES in the reaction. Finally, the absorption peaks observed at around 808 cm^{-1}, correspond to Si–C bonds. Si–C absorption peak indicates the presence of the CH$_3$SiO$_3$ groups on the surface of silica particles linked by condensation reaction of Si–OH, resulting in higher hydrophobicity [13].

3.2. Morphological Study

The representative images of the coatings deposited on glass substrates and treated at 150 °C and 400 °C are shown in Figure 3.

As expected, by comparing images for three cases, when the silica solution is deposited four times, a darker layer of particles is formed. It can be seen how the layer obtained for the case of the sol aged for 9 days, is more heterogeneous, showing zones with a higher concentration of silica particles because the degree of agglomeration of these particles increases with aging period [31–33].

Using ammonia, during the hydrolysis and condensation of TEOS, silica nanoparticles (with several Si–OH) progressively generate. According with other works [34,35], the hydrolysed MTES monomers cross-condense with Si–OH groups on the surface of silica particles, and then, link them into the porous network.

The Figures 4 and 5 show the SEM micrographs of the silica coatings with different dip-times and aging periods, respectively, all of them having been thermally treated at 400 °C. These reveal the size and shape of the prepared spherical silica particles, with the size of approximately 530 nm being uniform.

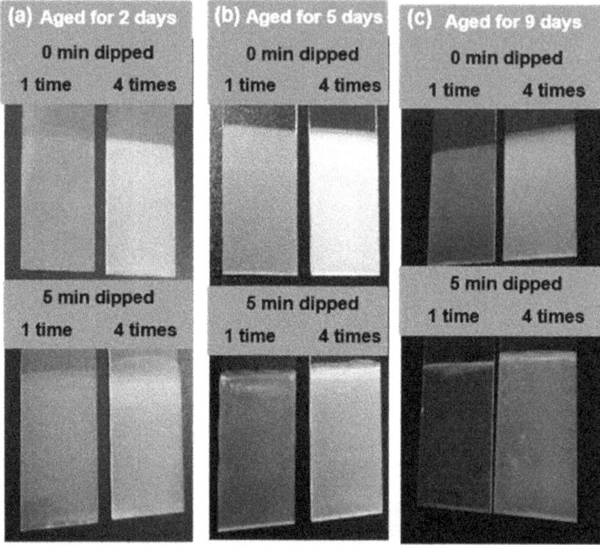

Figure 3. Representative images of the coatings aged for (**a**) 2, (**b**) 5 and (**c**) 9 days.

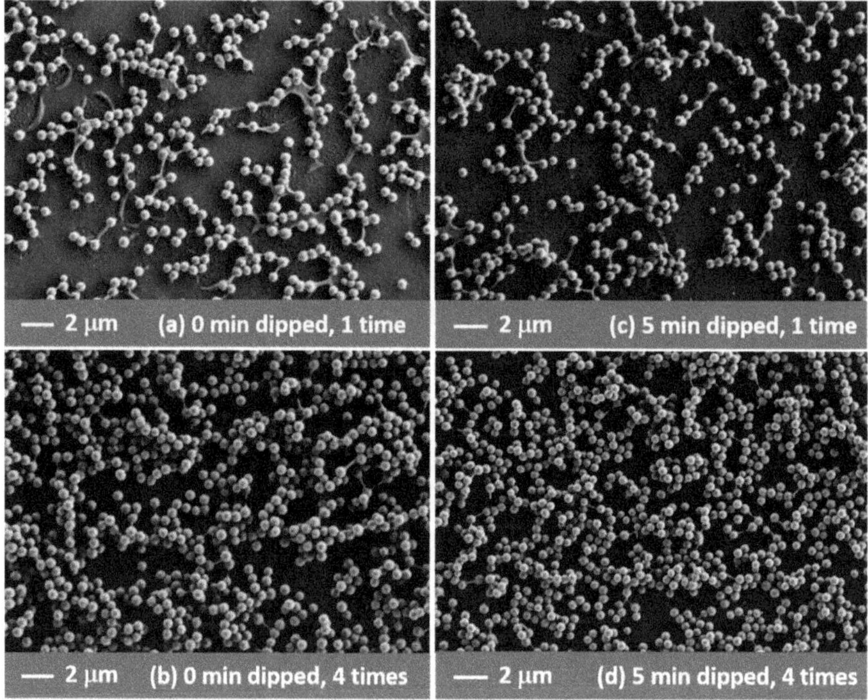

Figure 4. SEM images of hydrophobic silica coatings aged for 5 days, varying the dipping conditions: (**a**) dipped for 0 min once, (**b**) dipped for 0 min four times, (**c**) dipped for 5 min once and (**d**) dipped for 5 min four times.

As can be seen, the surface morphologies keep nearly the same after being immersed for 0 and 5 min. As the immersion times increase, however, the particle concentration increases, enhancing the surface roughness of the silica coating. In addition, after the aging, the SiO_2 particles begin to assemble, forming a nano-micro-binary structure (Figure 5) due to the condensation reaction of OH–group between SiO_2 particles [20].

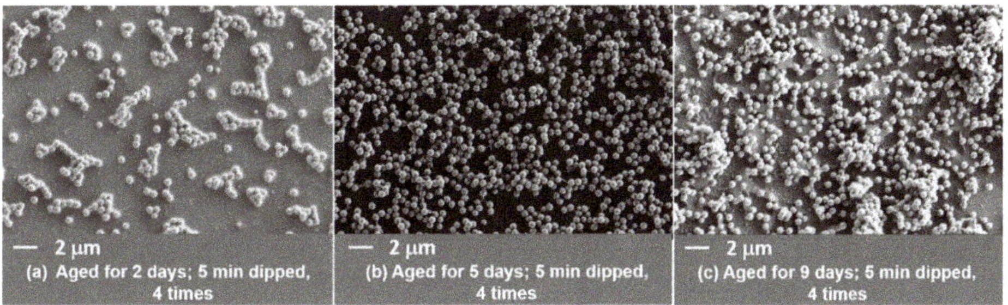

Figure 5. SEM images of hydrophobic silica coatings dipped for 5 min four times and aged for (**a**) 2 days, (**b**) 5 days and (**c**) 9 days.

However, the layer obtained by aging the sol for 9 days shows a surface more heterogeneous (Figure 3), being a handicap for its industrial application. Then, in order to improve the hydrophobic properties of the homogeneous obtained coatings, the coating aged for 5 days and with four immersions for 5 min was selected to study the effect of parameters such as dipping rate (and then, the thickness) on the hydrophobic properties.

Therefore, for this case, the glass substrates were dipped using faster dipping rates (500 and 1000 mm/min) for 5 min (four times). As is well known, the coating thickness increases with the coating cycle number and with the substrate velocity. Values of 520, 605–1150 and 774–1105 nm were obtained for the measured thickness of the coating using dipping rates of 200, 500 and 1000 mm/min, respectively. Therefore, in agreement with literature, the average coating thickness increases with increasing dipping rate [36–38].

The SEM images obtained for theses coatings are shown in Figure 6. Using dipping rates higher than 200, in this case 500 and 1000 mm/min, as it is observed for more aging days, there is a condensation reaction of OH– group between SiO_2 particles giving rise to a nano-micro-binary structure. Nevertheless, the coating shows a homogeneous distribution of particles (Figure 7).

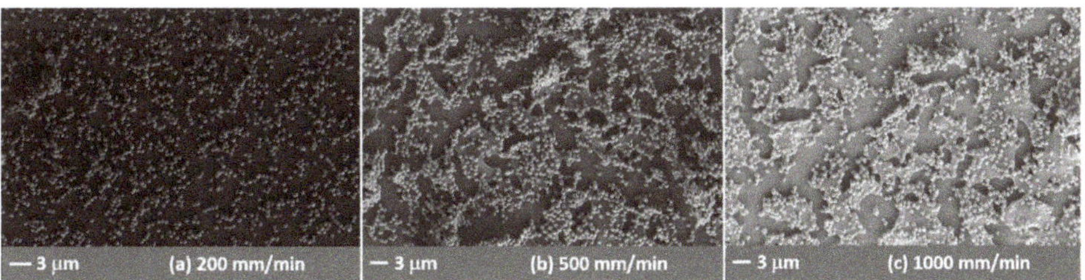

Figure 6. SEM images of hydrophobic silica coatings dipped for 5 min four times, and aged for 5 days with a dipping rate of (**a**) 200 mm/min, (**b**) 500 mm/min and (**c**) 1000 mm/min.

Figure 7. Representative appearance image of the coating aged for 5 days, obtained using a dipping velocity of 1000 mm/min.

The remaining SEM images, corresponding to the samples calcined at 150 and 400, 450, 500 and 550 °C are shown in Figure 8.

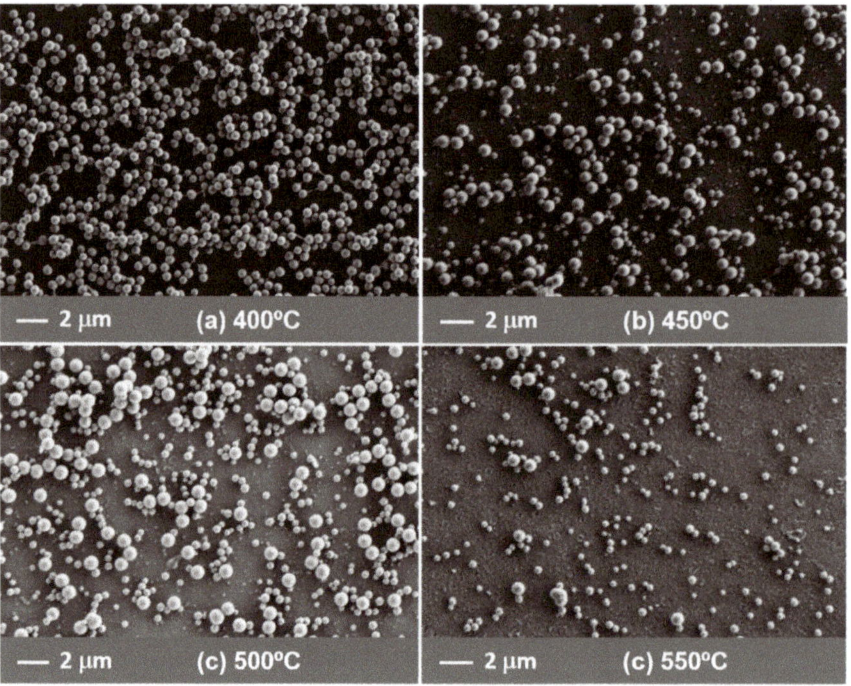

Figure 8. SEM images of silica coatings aged for 5 days and dipped for 5 min four times calcined at (**a**) 400, (**b**) 450, (**c**) 500 and (**d**) 550 °C.

The SEM images show that for calcination temperatures higher than 400 °C, the samples do not have a uniform size, with particle diameters being approximately, 220–710, 300–675 and 365–820 nm, for the samples calcined at 450, 500 and 550 °C, respectively. Then, as the temperature value of the thermal treatment increases, the size distribution is more heterogeneous. In addition, as can be observed, the particle concentration decreases slightly with the calcination temperature, which is caused by the adsorption between hydroxyl groups on the surface of SiO_2 particles [39].

3.3. Static and Sliding Water Contact Angle

The wetting behaviour of superhydrophobic surfaces depends on their chemical composition and surface microstructure.

The hydrophobic behaviour of silica coatings deposited on glass slides as a function of the duration of the dipping and the number of coatings was analysed by water contact angle (WCA). In all the cases the dipping speed was of 200 mm/min. As is well known, when a water droplet contacts a porous silica coating and spreads flat instantaneously on the surface, this coating shows a great superhydrophilic nature. The images of the static water contact angles measured for each experimental condition, are shown in Figure 9.

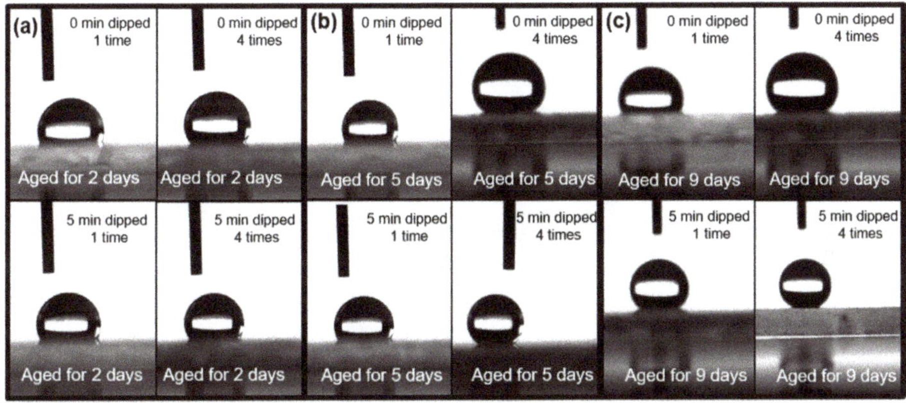

Figure 9. Water drop images on the coatings aged for (**a**) 2, (**b**) 5 and (**c**) 9 days.

The Table 1 lists the values of the static water contact angles and sliding angles (SA) obtained for all samples as a function of the duration of the dipping and the number of coatings.

Table 1. Values of water contact angle (WCA) and sliding angle (SA) obtained for samples as a function of the duration of the dipping and the number of coatings.

	Aged for 2 Days				Aged for 5 Days				Aged for 9 Days			
	0 min		5 min		0 min		5 min		0 min		5 min	
	WCA (°)	SA (°)	WCA (°)	SA (°)	WCA (°)	SA (°)	WCA (°)	SA (°)	WCA (°)	SA (°)	WCA (°)	SA (°)
1 time	108 ± 1	20 ± 1	115 ± 1	40 ± 1	116 ± 1	65 ± 2	117 ± 2	58 ± 2	118 ± 1	33 ± 1	120 ± 1	28 ± 1
4 times	123 ± 1	53 ± 1	121 ± 1	51 ± 2	124 ± 1	41 ± 2	132 ± 1	35 ± 1	137 ± 1	31 ± 1	149 ± 1	<5

As can be observed, the coatings obtained by dipping one time exhibit smaller values of contact angle than the obtained dipping four times for the three cases, showing that the number of times that the substrate is immersed has an important influence on the hydrophobic surface properties of the material under study. This behaviour can be explained due to the increase of the particle concentration with the immersion times (Figure 5) that enhances the surface roughness. Adding to this, an increase of the contact angle values was obtained by increasing the time during the substrate is dipped, especially for the samples aged for 5 and 9 days with four immersions, even though SEM did not show a significant difference of the particle concentration on the substrate surface (Figure 4). These results are in good agreement with the results obtained by H. Yang et al. [33].

In this study, the superhydrophobicity was reached when the number of assemblies increased up to four cycles, showing a water contact angle of 149° ± 1° for the sol aged for 9 days and a sliding angle of nearly <5°. This could be attributed to the increase of surface roughness [40]. The deposition of the aggregated silica microspheres on the glass substrates provides a rough surface, and the grooves between particles and surface roughness increases with the number of deposition cycles. These rough structures of the films allow them to trap enough air in the grooves between particles, giving rise to a repellence of water.

On the other hand, when the coating thickness increases, the surface roughness and the hydrophobicity increase, but the transparency and mechanical stability decrease. In addition, as has been above mentioned, the layer obtained by aging the sol for 9 days showed a surface more heterogeneous.

Figure 10 shows the water contact angles measured for the sample aged for 5 days with four rounds of dipping using a dipping rate of 1000 mm/min (cured at 400 °C).

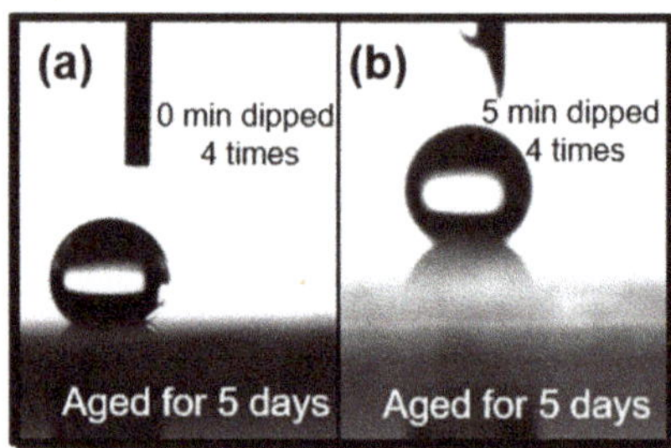

Figure 10. Water drops' images on the coatings using a dipping rate of 1000 mm/min.

The Table 2 lists the values of the water contact angles obtained for the coatings aged for 5 days and dipped for 5 min four times for the samples using different dipping rates. As is shown, using higher ripping rates, higher values of the water contact angles were obtained.

Table 2. Values of the WCA and SA obtained for the coatings aged for 5 days and dipped for 5 min times; samples used different dipping rates.

	200 mm/min		500 mm/min		1000 mm/min	
	WCA (°)	SA(°)	WCA (°)	SA(°)	WCA (°)	SA(°)
1 time	117 ± 1	58 ± 2	139 ± 1	10 ± 1	143 ± 2	<5
4 times	131 ± 1	35 ± 1	145 ± 1	<5	149 ± 1	<5

The above results suggested that the aging period, the deposition cycles, the time during the sample is dipped and the velocity used in the dip-coating process have great effects on the properties of the silica coatings that are critical for their desired application. The best results were obtained for the coating aged for 9 days and dipped for 5 min 4 times using a dipping rate of 200 mm/min (Table 1 and Figure 4c) and for the aged for 5 days and dipped for 5 min 4 times using a dipping rate of 1000 mm/min (Table 2 and Figure 5c). However, this last coating presents a superficial morphology more homogeneous.

3.4. Thermal Stability

The effect of the calcination temperature was confirmed by putting the SiO_2 superhydrophobic coating sample in an oven at 400, 450, 500 and 550 °C for 2 h, and then, it was tested by measuring the WCAs. The coatings were obtained by using a dipping speed of 1000 mm/min for those aged for 5 days and dipped four times.

The test results are shown in Figure 11, showing that above 400 °C the hydrophobicity decreases as coating curing temperature increases. In fact, when a calcination temperature of 550 °C is applied, the WCA decreased to nearly 0°, indicating that the wetting behaviour of the surface changed from superhydrophobicity to superhydrophilicity. This phenomenon, shown at relatively low temperatures (<600 °C) in other works about siloxane surfaces, was analysed in detail by Karapanagiotis et al., by studying the effects of thermal treatment temperature and time [41].

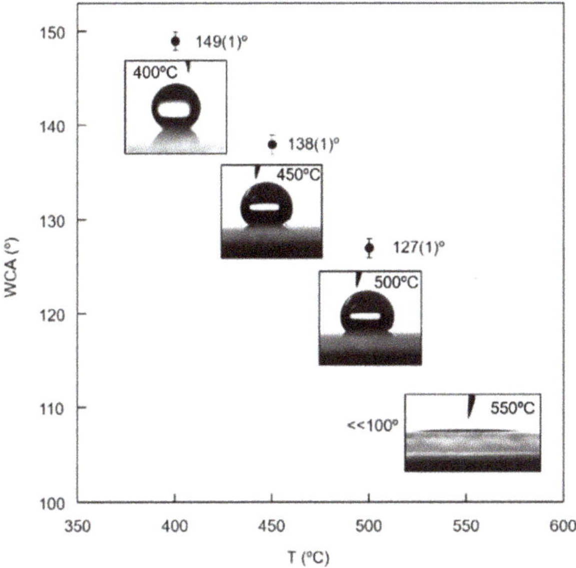

Figure 11. Water contact angle of the coating aged for 5 days and dipped four times at different curing temperatures.

That phenomenon takes place because the $-CH_3$ groups on the surface of SiO_2 particles are replaced by $-OH$ groups due to the decomposition of $-CH_3$ groups after calcining at a high temperature [34,41–43].

3.5. Thermal Degradation of the Superhydrophobic Coating

The thermogravimetric analysis (TGA) curves for the coatings aged for 5 days and cured at different temperatures are shown in Figure 12. The TGA curves had a main weight loss below 200 °C due to the physically adsorbed water molecules of the environmental humidity. After that, in the coating materials cured at 400 °C and 450 °C, the weight decreased gradually up to 750 °C, by means of the degradation of organic groups. For the coatings cured at 500 °C the weight loss curve is flat up to 600 °C, due to the fact that methyl groups were degraded during the thermal treatment of the coating. The total weight loss for the coatings cured at 400 °C and 450 °C is at about 8%, and for the coatings cured at 500 °C and 550 °C, is less than 5%, which indicates that they are thermally stable and inorganic.

For the sample cured at 400 °C, the weight-loss phenomena with respect to the increase in the temperature up to 500 °C corresponds to the degradation of $-CH_3$, and the SiO_2 coating finally develops a superhydrophilic nature. This behaviour has been previously observed, showing a WCA near to 0°. As can be observed, as the curing temperature increases, the silica products show less loss of mass because they have already lost mass during the curing heating.

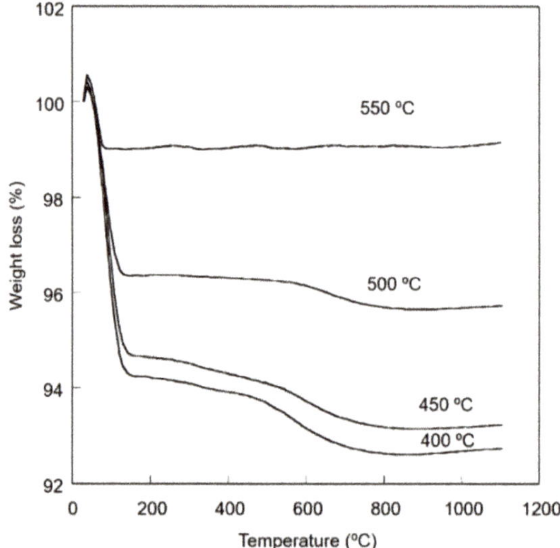

Figure 12. Thermogravimetric analysis (TGA) curves of the SiO_2 particles for a coating aged for 5 days and cured at 400, 450, 500 and 550 °C.

4. Conclusions

The superhydrophobic SiO_2 surface was successfully prepared with TEOS and MTES as precursors by a sol-gel method. The effect of the number of sol-gel coatings, the aging period and the dipping rate on the hydrophobic properties and thermal stability were analysed. Homogeneous silica coatings with static water contact angles as high as 149° were obtained for the sample aged for 5 days, dipped four times with a dipping rate of 1000 mm/min and cured at 400 °C. These thermally stable superhydrophobic coatings are potentially suitable for various industrial applications.

Author Contributions: Conceptualization, K.V., E.G., A.M.G, A.A.-I. and E.A.; methodology, K.V., E.G., A.M.G, A.A.-I. and E.A.; writing—review and editing, K.V., E.G., A.M.G., A.A.-I. and E.A.; supervision, E.G., A.M.G. and E.A.

Funding: This research received no external funding.

Acknowledgments: This work under "Multifunctional surfaces at the frontier of knowledge" project is supported by the Basque Government Industry Department under the ELKARTEK Programme.

Conflicts of Interest: The authors declare no conflict of interest.

References

1. Shahabadi, S.M.S.; Brant, J.A. Bio-inspired superhydrophobic and superoleophilic nanofibrous membranes for non-aqueous solvent and oil separation from water. *Sep. Purif. Technol.* **2019**, *210*, 587–599. [CrossRef]

2. Kim, W.; Kim, D.; Park, S.; Lee, D.; Hyun, H.; Kim, J. Engineering lotus leaf-inspired micro- and nanostructures for the manipulation of functional engineering. *J. Ind. Eng. Chem.* **2018**, *61*, 39–52. [CrossRef]

3. Xue, C.-H.; Jia, S.-T.; Zhang, J.; Ma, J.-Z. Large-area fabrication of superhydrophobic surfaces for practical applications: An overview. *Sci. Technol. Adv. Mater.* **2010**, *11*, 033002–033017. [CrossRef] [PubMed]

4. Ganbavl, V.V.; Bangi, U.K.H.; Latthe, S.S.; Mahadik, S.A.; Rao, A.V. Self-cleaning silica coatings on glass by single step sol–gel route. *Surf. Coat. Technol.* **2011**, *205*, 53338–55344. [CrossRef]

5. Mozumder, M.S.; Mourad, A.-H.I.; Pervez, H.; Surkatti, R. Recent developments in multifunctional coatings for solar panel applications: A review. *Sol. Energy Mater. Sol. Cells* **2019**, *189*, 75–102. [CrossRef]

6. Yang, M.; Liu, W.; Jiang, C.; He, S.; Xie, Y.; Wang, Z. Fabrication of superhydrophobic cotton fabric with fluorinated TiO_2 sol by a green and one-step sol-gel process. *Carbohydr. Polym.* **2018**, *197*, 75–82. [CrossRef]

7. Ma, M.; Hill, R.M. Superhydrophobic surfaces. *Curr. Opin. Colloid Interface Sci.* **2006**, *11*, 193–202. [CrossRef]

8. Fihri, A.; Bovero, E.; Al-Shahrani, A.; Al-Ghamdi, A.; Alabedi, G. Recent progress in superhydrophobic coatings used for steel protection: A review. *Colloids Surf. A Physicochem. Eng. Asp.* **2017**, *520*, 378–390. [CrossRef]

9. Seraji, M.M.; Sameri, G.; Davarparah, J.; Bahramian, A.R. The effect of high temperature sol-gel polymerization parameters on the microstructure and properties of hydrophobic phenol-formaldehyde/silica hybrid aerogels. *J. Colloid Interface Sci.* **2017**, *493*, 103–110. [CrossRef]

10. Cui, S.; Liu, Y.; Fan, M.-H.; Coopere, A.T.; Liu, X.-Y.; Han, G.-F.; Shen, X.-D. Temperature dependent microstructure of MTES modified hydrophobic silica aerogels. *Mater. Lett.* **2011**, *65*, 606–609. [CrossRef]

11. Izarra, I.; Cubillo, J.; Serrano, A.; Rodríguez, J.F.; Carmona, M. A hydrophobic release agent containing SiO2-CH3 submicron-sized particles for water proofing mortar structures. *Constr. Build. Mater.* **2019**, *199*, 30–39. [CrossRef]

12. Sun, Z.; Liu, B.; Huang, S.; Wu, J.; Zhang, Q. Facile fabrication of superhydrophobic coating based on polysiloxane emulsion. *Progr. Org. Coat.* **2017**, *102*, 131–137. [CrossRef]

13. Latthea, S.S.; Imai, H.; Ganesan, V.; Rao, A.V. Porous superhydrophobic silica films by sol–gel process. *Microporous Mesoporous Mater.* **2010**, *130*, 115–121. [CrossRef]

14. Sheen, Y.-C.; Chang, W.-H.; Chen, W.-C.; Chang, Y.-H.; Huang, Y.-C.; Chang, F.-C. Non-fluorinated superamphiphobic surfaces through sol–gel processing of methyltriethoxysilane and tetraethoxysilane. *Mater. Chem. Phys.* **2009**, *114*, 63–68. [CrossRef]

15. Latthe, S.S.; Imai, H.; Ganesan, V.; Rao, A.V. Superhydrophobic silica films by sol-gel co-precursor method. *Appl. Surf. Sci.* **2009**, *256*, 217–222. [CrossRef]

16. Mahadik, S.A.; Kavale, M.S.; Mukherjee, S.K.; Rao, A.V. Transparent Superhydrophobic silica coatings on glass by sol–gel method. *Appl. Surf. Sci.* **2010**, *257*, 333–339. [CrossRef]

17. Shao, Z.; Luo, F.; Cheng, X.; Zhang, Y. Superhydrophobic sodium silicate based silica aerogel prepared by ambient pressure drying. *Mater. Chem. Phys.* **2010**, *257*, 333–339. [CrossRef]

18. Zhao, A.; Li, Y.; Li, B.; Hu, T.; Yang, Y.; Li, L.; Zhang, J. Environmentally benign and durable superhydrophobic coatings based on SiO2 nanoparticles and silanes. *J. Colloid Interface Sci.* **2019**, *542*, 8–14. [CrossRef]

19. Zheng, X.; Fu, S. Reconstructing micro/nano hierarchical structures particle with nanocellulose for superhydrophobic coatings. *Colloids Surf. A Physicochem. Eng. Asp.* **2019**, *560*, 171–179. [CrossRef]

20. Guilong, X.; Pihui, P.; Ermei, H.; Xiufang, W.; Dafeng, Z.; Zhuoru, Y. Preparation and characterization of superhydrophobic/superoleophilic silica film. *J. Chim. Ceram. Soc.* **2011**, *39*, 854–858. [CrossRef]

21. Jiang, Z.; Fang, S.; Wang, C.; Wang, H.; Ji, C. Durable polyorganosiloxane superhydrophobic films with a hierarchical structure by sol-gel and heat treatment method. *Appl. Surf. Sci.* **2016**, *390*, 993–10001. [CrossRef]

22. Zhang, Y.; Dong, B.; Wang, S.; Zhauo, L.; Wan, L.; Wang, E. Mechanically robust, thermally stable, highly transparent superhydrophobic coating with low-temperature sol–gel process. *RSC adv.* **2017**, *7*, 47357–47365. [CrossRef]

23. Rao, A.V.; Latthe, S.S.; Mahadik, S.A.; Kappenstein, C. Mechanically stable and corrosion resistant superhydrophobic sol–gel coatings on copper substrate. *Appl. Surf. Sci.* **2011**, *257*, 5442–5776. [CrossRef]

24. Xu, J.; Liu, Y.; Du, W.; Lei, W.; Si, X.; Zhou, T.; Lin, J.; Peng, L. Superhydrophobic silica antireflective coatings with high transmittance via one-step sol-gel process. *Thin Solid Films* **2017**, *631*, 193–199. [CrossRef]

25. Nguyen-Tria, P.; Tran, H.N.; Plamondon, C.O.; Tuduri, L.; Vo, D.-V.N.; Nanda, S.; Mishra, A.; Chao, H.-P.; Bajpai, A.K. Recent progress in the preparation, properties and applications of superhydrophobic nano-based coatings and surfaces: A review. *Prog. Org. Coat.* **2019**, *132*, 235–256. [CrossRef]

26. Pantoja, M.; Velasco, F.; Abenojar, J.; Martinez, M.A. Development of superhydrophobic coatings on AISI 304 austenitic stainless steel with different surface pretreatments. *Thin Solid Films* **2019**, *671*, 22–30. [CrossRef]

27. Jeevahan, J.; Chandrasekaran, M.; Joseph, G.B.; Durairaj, R.B.; Mageshwaran, G. Superhydrophobic surfaces: A review on fundamentals, applications, and challenges. *J. Coat. Technol. Res.* **2018**, *15*, 231–250. [CrossRef]

28. Stöber, W.; Fink, A.; Bohn, E. Controlled growth of monodisperse silica spheres in the micron size range. *J. Colloid Interface Sci.* **1968**, *26*, 62–69. [CrossRef]

29. Criado, M.; Sobrados, I.; Sanz, J. Polymerization of hybrid organic–inorganic materials from several silicon compounds followed by TGA/DTA, FTIR and NMR techniques. *Progr. Org. Coat.* **2014**, *77*, 880–891. [CrossRef]

30. Montejo, M.; Partal Ureña, F.; Márquez, F.; Ignatyev, I.S.; López González, J.J. Vibrational spectrum of methoxytrimethylsilane. *J. Mol. Struct.* **2005**, *744–747*, 331–338. [CrossRef]

31. Innocenzi, P. Infrared spectroscopy of sol–gel derived silica-based films: A spectra-microstructure overview. *J. Non-Cryst. Solids* **2003**, *316*, 309–319. [CrossRef]

32. Azlina, H.N.; Hasnidawani, J.N.; Norita, H.; Surip, S.N. Synthesis of SiO2 Nanostructures Using Sol-Gel Method. *Acta Phys. Pol. A* **2016**, *129*, 842–844. [CrossRef]

33. Yang, H.; Pi, P.; Cai, Z.-Q.; Wen, X.; Wang, X.; Cheng, J.; Yang, Z.-R. Facile preparation of super-hydrophobic and super-oleophilic silica film on stainless steel mesh via sol-gel process. *Appl. Surf. Sci.* **2010**, *256*, 4095–4102. [CrossRef]

34. Zhong, M.; Zhang, Y.; Li, X.; Wu, X. Facile fabrication of durable superhydrophobic silica/epoxy resin coatings with compatible transparency and stability. *Surf. Coat. Technol.* **2018**, *347*, 191–198. [CrossRef]

35. Wen, X.-F.; Wang, K.; Pi, P.-H.; Yang, J.-X.; Cai, Z.-Q.; Zhang, L.-J.; Qian, Y.; Yang, Z.-R.; Zheng, D.-F.; Cheng, J. Organic–inorganic hybrid superhydrophobic surfaces using methyltriethoxysilane and tetraethoxysilane sol–gel derived materials in emulsion. *Appl. Surf. Sci.* **2011**, *258*, 991–998. [CrossRef]

36. Strawbridge, I.; James, P.F. The factors affecting the thickness of sol-gel derived silica coatings prepared by dipping. *J. Non-Cryst. Solids* **1986**, *86*, 381–393. [CrossRef]

37. Guglielmi, M.; Zenezini, S. The thickness of sol-gel silica coatings obtained by dipping. *J. Non-Cryst. Solids* **1990**, *121*, 303–309. [CrossRef]

38. Brinfer, C.J.; Frye, G.C.; Hurd, A.J.; Ashley, C.S. Fundamentals of sol-gel dip coating. *Thin Solid Films* **1991**, *201*, 97–108. [CrossRef]

39. Li, K.; Zeng, X.; Li, H.; Lai, X.; Xie, H. Effects of calcination temperature on the microstructure and wetting behavior of superhydrophobic polydimethylsiloxane/silica coating. *Colloids Surf. A Physicochem. Eng. Asp.* **2014**, *445*, 111–118. [CrossRef]

40. Shang, Q.; Zhou, Y. Fabrication of transparent superhydrophobic porous silica coating for self-cleaning and anti-fogging. *Ceram. Int.* **2016**, *42*, 8706–8871. [CrossRef]

41. Karapanagiotis, I.; Manoudis, P.; Zurba, A.; Lampakis, D. From hydrophobic to superhydrophobic and superhydrophilic siloxanes by thermal treatment. *Langmuir* **2014**, *30*, 13235–13243. [CrossRef] [PubMed]

42. Yang, H.; Zhang, X.; Cai, Z.-Q.; Pi, P.; Zheng, D.; Wen, X.; Cheng, J.; Yang, Z.-R. Functional silica film on stainless steel mesh with tunable wettability. *Surf. Coat. Technol.* **2011**, *205*, 5387–5393. [CrossRef]

43. Kuo, C.Y.; Gau, C. Control of Superhydrophilicity and Superhydrophobicity of a Superwetting Silicon Nanowire Surface. *J. Electrochem. Soc.* **2010**, *157*, k201–k205. [CrossRef]

 *coatings*

Article

HiPIMS and DC Magnetron Sputter-Coated Silver Films for High-Temperature Durable Reflectors

Sophie Gledhill *, Kevin Steyer, Charlotte Weiss and Christina Hildebrandt

Fraunhofer Institute for Solar Energy Systems, Heidenhofstraße 2, 79110 Freiburg, Germany;
kevin.steyer@ise.fraunhofer.de (K.S.); charlotte.weiss@ise.fraunhofer.de (C.W.);
christina.hildebrandt@ise.fraunhofer.de (C.H.)
* Correspondence: sophie.gledhill@ise.fraunhofer.de; Tel.: +49-761-4558-2049

Received: 10 July 2019; Accepted: 17 September 2019; Published: 20 September 2019

Abstract: High-temperature durable mirrors based on a protected silver sputter coating are attractive for secondary reflector applications in concentrated solar thermal power plants. In this paper, silver films are deposited by high-power impulse magnetron sputtering (HiPIMS) and standard direct current (DC) magnetron sputtering, either as exposed discretely deposited films or in-sequence-deposited thin film systems, where the silver is protected and embedded between adhesion and barrier layers. The unprotected silver films and equivalent protected silver thin film systems are compared and characterized as deposited and after 400 °C oven temperature exposure. The reflectance is measured and grazing incident X-ray diffraction (GIXRD) and scanning electron microscopy (SEM) pictures were taken. The HiPIMS silver film, sputtered with a peak current of 200 A and an approximately equivalent average power density to the DC magnetron sputtered silver, exhibits higher reflectance (and conductivity). Increasing the power density further, yields silver films with lower reflectance, correlating to a reduced grain size. In the protected silver film system, the reflectance does not improve, due to the presence of a less reflective top adhesion layer. The protected film system, with the 200 A HiPIMS, is, however, more durable at 400 °C than the DC magnetron sputtered equivalent.

Keywords: protected silver; secondary reflector; sputtering; HiPIMS; GIXRD; thin film

---

1. Introduction

Concentrated solar power (CSP) systems generate solar power by using mirrors or lenses (primary concentrators or reflectors) to concentrate a large area of sunlight, or solar thermal energy, onto a smaller area (the receiver). A secondary concentrator reflects the sunlight coming from the primary concentrators onto the absorbing receiver. They ensure the gathering and redirecting, and in some applications further concentration and focusing, of the solar beams towards the absorber. A good overview of the available different secondary reflector material systems and different CSP configurations is given in [1].

A secondary reflector which is installed adjacent to the solar tower or linear Fresnel collector receiver can improve the optical efficiency of the system considerably as well as reducing its cost by [2]:

- Reducing the amount of flux which is spilled-off the target
- Improving the flux distribution uniformity
- Decreasing heat losses and CAPEX (capital expenditure) by reducing the receiver size

The main challenge for such a mirror is its high operation temperature. For the solar power tower, the high temperatures evolve mainly due to the amount of irradiance that is being absorbed by the mirror, and to a lesser extent due to convective heat exchange with its surrounding. One possible approach is to water-cool the mirror. However, this solution will mean an investment increase, technical challenges, and cause a plant shut-down in case of any failure of the cooling system.

A different approach avoiding water-cooling was to develop a protected silver coated secondary mirror which could retain its optical and mechanical properties up to 350 °C. The developed mirrors are based on a sputtered silver layer and several adhesion and barrier layers on a highly polished steel substrate. The silver layers are sputtered using either standard direct current magnetron sputtering (DC MS) or, alternatively, high-power impulse magnetron sputtering (HiPIMS) [3]. With HiPIMS, high power (up to megawatt range) is applied to the magnetron target in unipolar pulses at a low duty cycle and low repetition frequency while keeping the average power about two orders of magnitude lower than the peak power. This results in a high plasma density and high ionization fraction of the sputtered vapor. Previous studies with silver HiPIMS layers show an improvement in electrical conductance attributed to atomically smoother and denser films [4,5] being grown from the high energy flux of the sputtered silver.

For protected silver coatings for reflector applications, a desired improvement in both reflectance and film stability was the motivation to investigate HiPIMS silver in comparison to the standard DC MS silver. As the silver films must be durable at high temperature, the films were examined (optically and micro-structurally) as deposited and after exposure to a temperature of 400 °C (plausible operational conditions for the stringent solar power tower reflector). It should be noted that the protected silver coatings were tested under different conditions (e.g., a humidity and heat of 85% and 85 °C, respectively), but that is not the focus of the paper here. Other applications of protected sputtered silver where highly reflective yet highly durable reflector coatings are required, include telescopes (such as the coatings used at the Gemini Observatory telescope [6]) or in robust floodlight applications.

2. Materials and Methods

The HiPIMS and DC MS silver films were prepared in house at the Fraunhofer Insitute of Solar Energy Systems. Both processes used 99.99% pure silver targets which were sputtered in argon atmospheres of approximately 300 Pa. A TruPlasma Highpulse 4008 DC generator (Trumpf-Hüttinger, Freiburg, Germany) was used for the HiPIMS mode. The DC magnetron was powered by a TIG-DCS (Trumpf-Hüttinger, Freiburg, Germany). The electrical sputter parameters for the DC MS and HiPIMS silver are given in Table 1. For HiPIMS deposition, a pulse frequency of 150 s^{-1} and pulse duration of 80 µs was used.

Table 1. Electrical sputter parameters used to produce the silver layers. The parameters given in bold correlate with the conditions used to prepare both the discrete silver layer and the embedded silver film in the protected silver layer system. The approximate peak power density (P_{peak}) and the average power density (P_{average}) are given here. The dynamic deposition rate is given as is and normalized to the power density.

Target Power Delivery Mode	I (A) [1]	P_{peak} (W·cm^{-2})	P_{average} (W·cm^{-2})	Dynamic Deposition Rate (nm·(m min^{-1}))	Dynamic Power Density Normalized Deposition Rate (nm·(m min^{-1}) (kW m^{-2})$^{-1}$)
DC	12	4	4	66	1.65
HiPIMS	100	192	2.3	21	0.91
	200	320	3.8	27	0.71
	300	621	7.4	47	0.61
	400	1005	12.1	67	0.55

[1] This is the peak current for the HiPIMS, the constant current in DC MS mode.

A current of 12 A in the DC magnetron mode was chosen, as this was the standard silver used in the protected silver system. It exhibits the highest temperature durability for DC magnetron mode. The 200 A HiPIMS silver was deposited with a similar power density to the 12 A DC MS sample. The power density is given only as an approximate value, as it is calculated from the power and the approximate race-track area of the silver target magnetron. The measured dynamic rate, normalized to the power density, is given in Table 1. As typically seen, HiPIMS mode led to a drop in the deposition rate for equivalent power density. This is caused by increased film density, increased self-sputtering, and sputtering of the film at higher peak currents, as discussed in [7].

Cleaned float glass was used as a substrate for reflectance and conductivity measurements. Glass with a sputtered 45 nm NiCr (Ni:Cr 80:20 wt % target DC magnetron, 9.5 W/cm^2 in an Ar pressure of 0.16 Pa) adhesion layer was used as a substrate for the reflectance, grazing incidence X-ray diffraction (GIXRD), and scanning electron microscopy (SEM) analysis. It should be noted that for the SEM and GIXRD measurements, the adhesion layer coating which the silver is sputtered on was identical for both the discrete and the protected silver.

The protected silver layer system was deposited also on polished 1 mm 1.4301 steel substrates. The protected sputtered silver system was NiCr (45 nm)/Ag (120 nm)/NiCr (0.5 nm)/Si$_x$N$_y$ (100 nm) and is shown in Figure 1. The Si$_x$N$_y$ barrier and NiCr adhesion layers were sputtered in sequence with the silver films. The bottom NiCr layer was as described above. The top NiCr layer used a standard DC planar magnetron (target Ni:Cr 80:20 wt %) with a power of 0.8 W/cm^2 in an Ar atmosphere of pressure 0.09 Pa. The Si$_x$N$_y$ was deposited using a middle frequency, MF, dual rotary cathode (target Si:Al 90:10 wt %) at 10.8 W/cm^2 in an Ar/N2 (gas flow rate ratio 1:1) at a pressure of 0.3 Pa. The top adhesion was a nominal 0.5 nm NiCr designed to maximize the Ag-to-barrier layer adhesion with minimal reflectance degradation.

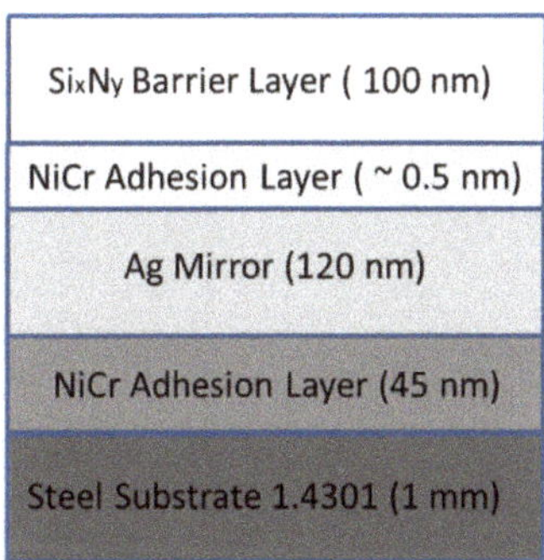

Figure 1. Schematic of the protected silver layer system.

All GIXRD patterns were recorded using a Philips X'Pert MRD system equipped with a CuK$_\alpha$ X-ray source. The Scherrer [8] formula,

$$\tau = \frac{K\lambda}{\beta \cos \theta},$$ (1)

was used as a comparative empirical guide (where τ is the grain size; K is a shape factor, here with a value of 0.9; λ is the X-ray wavelength of 1.54 Å; β is the full width half maximum; and θ is the Bragg angle), rather than an accurate calculation of the average grain size. The shifts in peak position with annealing were estimated and compared, again to determine a trend between samples rather than a precise measurement of strain in the silver crystal lattice.

The scanning electron microscope used was a FIB-SEM Auriga 60 (Zeiss, Oberkochen, Germany). Reflectance measurements were done with a Vertex 80 spectrometer (Bruker, Ettlingen, Germany) with an integrated sphere. The solar hemispherical reflectance is the percentage of the direct and diffuse AM1.5 solar radiance reflected from the coating system. Sheet resistance measurements were taken with a four-point probe (only for pure silver-coated glass substrates). The conductivity was thus calculated from the measured film thickness. Sputtered film thickness was checked using both a Dektak 6M profilmeter (Veeco, New York, NY, USA) and a calibrated Xray Fluorescence Spectrometer, XRF, Fischerscope XDV-µ (Helmut-Fischer, Sindelfingen, Germany).

3. Results

The results section is divided into two subsections for clarity. Firstly, the results for the discrete silver films are presented, and secondly, the results for the protected silver film system are given.

3.1. Discrete Silver Films

The discrete silver thin films were investigated in addition to the protected silver film system, as having no top coatings allows for a deeper understanding of the silver layers themselves.

The reflectance spectra of the as-deposited discrete 500 nm silver films on glass substrates are presented in Figure 2a. The solar reflectance increased to a maximum for the silver films deposited with 200 A in HiPIMS mode, but for higher HiPIMS currents it decreased and, indeed, with 400 A HiPIMS it was lower than the 12 A DC MS-deposited silver film. In Figure 2b, the conductivity of the films is plotted against the global solar hemispherical reflectance. The films sputtered by HiPIMS exhibit an approximately linear relationship whereby the conductivity proportionally increases with increasing reflectance. The 12 A DC MS samples are sputtered in a different mode and exhibit a higher conductivity-to-reflectance ratio than the HiPIMS films.

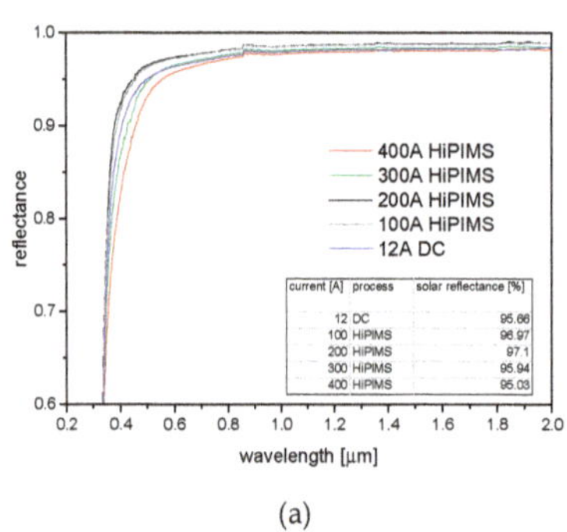
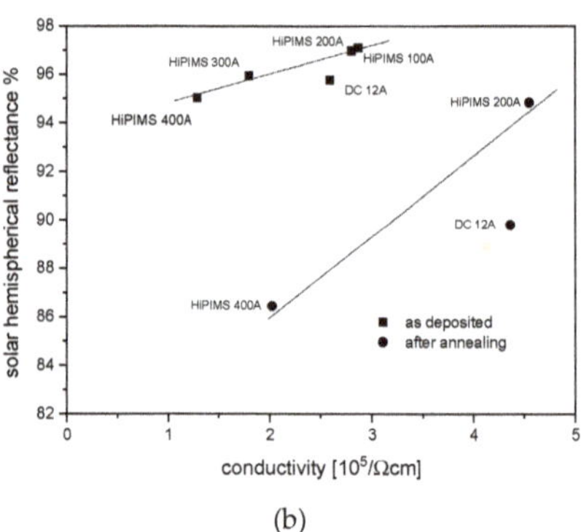

(a)   (b)

Figure 2. (**a**) Reflectance spectrum of the discrete 500 nm silver films sputtered on glass substrates with different currents (direct current (DC) and high-power impulse magnetron sputtering (HiPIMS) mode). The embedded table shows the global solar (AM 1.5 radiance) hemispherical reflectance for the silver films sputtered with different parameters. (**b**) Silver film conductivity against global solar hemispherical reflectance (500 nm silver on glass), both as deposited and after annealing for 8 h at 400 °C. The dotted lines serve only as a guide to the eye.

The films were annealed at 400 °C for 0.5, 4, 8, 16, and 32 h, the reflectance and conductivity were subsequently measured. With increasing annealing time, the conductivity of all the silver films increased, but solar reflectance decreased. What is more, the reflectance changed from being purely specular to partially diffuse. After 8 h annealing at 400 °C, corresponding to the data shown in Figure 2b, the 200 A HiPIMS silver films remained highly reflective (94% total) but the diffuse solar reflectance contribution was 40%, whilst the 12 A DC MS silver and the highly energetically sputtered 400 A HiPIMS silver had a lower total reflectance (89% and 84%, respectively) but also a lower diffuse contribution (10% and 2%, respectively). Due to this high optical scattering, the 200 A HiPIMS films appeared white after annealing.

SEM pictures are shown in Figure 3 of equivalent 120 nm silver films grown on the NiCr-coated glass. The as-deposited films appeared smooth and relatively featureless at low magnification. The grains, observable at high magnification, for the 12 A DC MS silver and the 200 A HiPIMS silver were apparently of a similar size (≤20 nm), whereas the 400 A HiPIMS grains were distinguishably smaller (≤10 nm). After annealing (again for 8 h at 400 °C), all samples exhibited the characteristic agglomeration (large crystals) and hole (void) formation of unprotected silver exposed to elevated temperatures reported in [9]. Moreover, the grain growth in the bulk of the film was apparent for all samples. The average bulk grains, post annealing, for the 200 A HiPIMS appeared the largest (50–150 nm), whilst the 400 A HiPIMS appeared the smallest (10–50 nm).

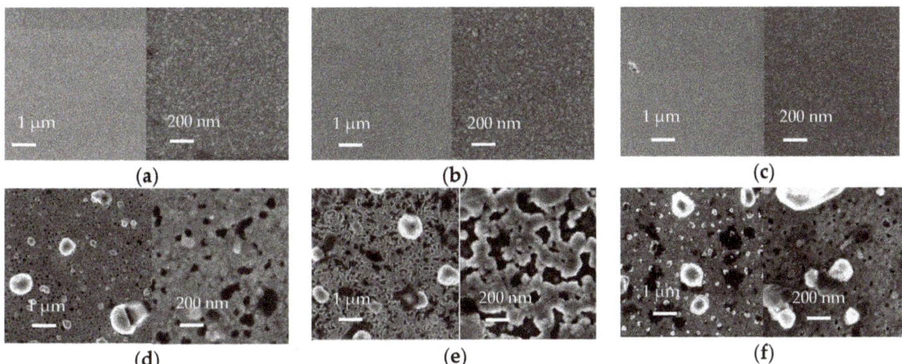

Figure 3. Scanning electron microscopy (SEM) pictures. Each picture pair is from the same sample, the one on the right being of higher magnification than the one on the left. The scale is given in (**a**). (**a–c**) The upper pictures show the films as deposited after sputtering, whilst (**d–f**) shows pictures post 8 h anneal at 400 °C; (**a,d**) shows the 12 A direct current magnetron sputtering (DC MS) silver; (**b,e**) shows the 200 A HiPIMS silver; and (**c,f**) shows the 400 A HiPIMS silver.

It is clear from the SEM pictures that the smooth silver films became significantly rougher, caused by the agglomeration and voiding [9,10]. This can be correlated with the decrease in spectral and total solar reflectance. The rougher surface features would cause the observed pronounced scattering despite the increased grain size, with annealing, which allows the conductivity to improve, provided the film remains networked.

In Figure 4a, the GIXRD spectra for the 400 A HiPIMS, 200 A HiPIMS, and 12 A DC MS silver films deposited on NiCr coated glass substrates, as-deposited and after an 8 h anneal at 400 °C, are shown. As stated, the grain size, peak position and intensity were calculated or measured from the spectra, and for clarity and transparency they are given in Table S1 as supplementary material.

Randomly oriented polycrystalline silver has a diffraction peak ratio of (111):(200):(220):(311):(222) as 100:40:25:26:12 [11]. In all as-deposited films sputtered here, the (220) plane exhibited a larger relative intensity than the random oriented reference.

After annealing, the diffraction from the (220) plane had the highest intensity, and the relative contribution from the (111) planes reduced significantly as the grains increased in orientation. This is particularly pronounced in the 200A HiPIMS silver film.

Using the Scherrer formula as a qualitative indication of grain size, it is implicated that for the as-deposited samples, the 200 A HiPIMS silver film has marginally larger grains than the 12 A DC MS silver film, but the 400 A HiPIMS silver film has grains smaller in size. This correlates with what is seen in the SEM pictures. It can be also correlated to the trend in conductivity of the silver film on the glass substrate: a smaller grain size yields lower sheet conductivity. There is no clear peak shift on annealing, although the observed narrowing of the peaks with annealing indicates grain growth in the sample, again in correlation with the SEM pictures and the increase in conductivity with annealing in equivalent films. Figure 4b is discussed in the context of the next section.

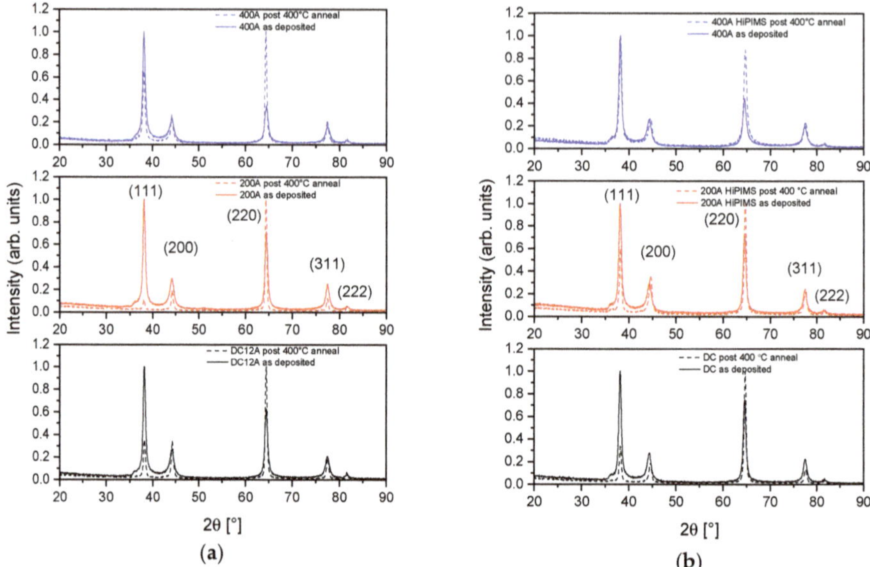

Figure 4. Grazing incident X-ray diffraction (GIXRD) for the 400 A HiPIMS, 200 A HiPIMS, and 12 A DC MS 120 nm sputtered silver (**a**) as discrete films on coated-glass substrates as deposited and after 8 h anneal at 400 °C; (**b**) as protected silver coating systems on glass substrates, as deposited (solid line), and after 24 h anneal at 400 °C. The adhesion layer underneath the silver is identical in both cases for the GIXRD measurements.

3.2. Protected Silver Thin Film System

In this section, the results of the protected silver thin film systems will be presented.

GIXRD spectra shown in Figure 4b for the 400 A HiPIMS, 200 A HiPIMS, and 12 A DC MS protected silver film system deposited on coated glass substrates are shown as deposited and after a 24 h anneal at 400 °C. Glass substrates were used for the GIXRD to avoid any interference diffraction peaks from the underlying steel substrate. Both the discrete silver discussed earlier and the protected silver system were grown on the same 45 nm NiCr adhesion layer coating the glass. The difference was that the protected silver system had a top NiCr adhesion and Si_xN_y barrier coating. As previously, the grain size, peak position, and intensity were calculated or measured from the spectra, and for clarity and transparency they are given in Table S2 as supplementary material. The as-deposited protected silver films exhibit the same trend in grain sizes as the as-deposited discrete silver films, and the grains are likewise orientated. After the 24 h anneal at 400 °C, the grains were larger and became even more aligned, with the (111) peak reducing and the (220) peak increasing in amplitude. What is different from the discrete silver films is that after annealing the protected silver films, there was a clear peak shift to a higher Bragg angle. The average peak shift was higher in the HiPIMS films than in the DC MS film. This is an indication that the silver protected film is compressively strained after the annealing step.

Figure 5a shows the global reflectance of the as-deposited protected silver thin films systems. The total solar reflectance of all the film systems was reduced to below 90%. In this case here, the top adhesion layer was an ultra-thin NiCr deposition (nominally 0.5 nm thick, in practice an incomplete covering). Adding additional layers, such as the Si_xN_y barrier layer, on top of the silver film reduced the total reflectance due to unwanted interference from additional interfacial reflectance and absorption by the adhesion and barrier layer. This was particularly prevalent in the UV-visible portion of the spectrum. Due to this, the gain in reflectance that was achieved for the discrete silver by depositing with the 200 A HiPIMS instead of 12 A DC MS silver was not apparent. The reflectance of the film system incorporating the 400 A HiPIMS silver was however lower, which was most likely due to the poorer silver reflectance, since the film system was identical between samples.

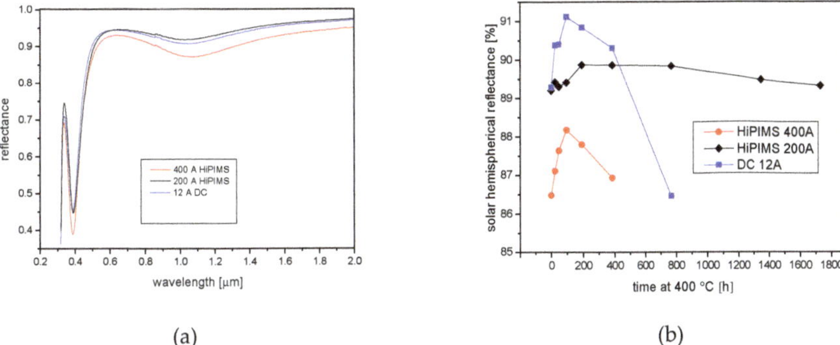

(a) (b)

Figure 5. (**a**) Reflectance spectrum of the protected silver layer system (as deposited) on steel substrates for the three alternative silver deposition conditions; (**b**) Global solar hemispherical reflectance plotted against time in the oven at 400 °C.

The protected silver thin film systems were subsequently annealed at two-fold increasing intervals, e.g., 24, 48, 96, 192, and 384 h, until the sample was deemed to have degraded, either due to degradation apparent to the eye or a loss in optical reflectance below the starting value. The silver, being embedded in the adhesion and barrier layers, was protected from corroding elements in the atmosphere as well as agglomeration. Agglomeration, which began instantaneously at 400 °C for the discrete silver films, was limited in the protected silver system by the adhesion layers and barrier layers. For the first 96 h of oven time, all the silver film systems increased in reflectance. This can be partially attributed to grain growth in the silver. This occurred in the protected silver system without surface roughening due to pronounced agglomeration, which caused the optical scattering observed in the discrete silver films. The increase in reflectance may also be due to the oxidation of the top metallic adhesion layer above the silver, which subsequently exhibited a higher transmission, particularly in the UV-visible portion of the spectrum. The increase in the 12 A DC MS and 400 HiPIMS silver film systems on annealing was a 2% absolute rise in reflectance, whereas the 200 A HiPIMS silver system increased by less than 1%.

After more than 96 h of exposure, the reflectance either plateaued or decreased. The decrease in reflectance was because of localized agglomeration of the silver. The degradation with the temperature of the silver is dependent, not only on the sputtered silver layer but on the film system. The various degradation mechanisms of the protected silver film system are discussed elsewhere [12]. In this paper, the film systems were identical, except for the silver deposition conditions. Degradation around the edges of all samples was apparent (see Figure 6 for a photograph of the mirrors after maximal oven exposure) due to higher sideways ingression of oxygen or moisture from the sample edge. Localized discrete points of degradation appeared in the bulk of the film and expanded with increased time at 400 °C. This is more pronounced in the sample prepared with the 12 A DC MS silver. The SEM picture in Figure 7a shows a typical degradation patch from a protected 12 A DC MS silver system sample. The silver agglomerated causing rupturing of the top barrier layer coating and thus further localized accelerated agglomeration and thus larger degradation patches with time. The large crystal on the surface was silver. It was confirmed by Energy Dispersive X-ray analysis that the crystal was silver. The mechanism of corrosion was similar for the HiPIMS silver film systems. A typical example is shown in the SEM picture in Figure 7b, for the 200 A HiPIMS sample after 768 h at 400 °C. Agglomeration of the silver took place, as seen for the DC MS; however, the pronounced rupturing of the barrier layer was not observed; thus, the points of degradation did not expand at an accelerated rate and hence resulted in smaller degradation points.

(a)   (b)   (c)

Figure 6. Photographs of 5 × 5 cm protected silver coated steel after end of testing in oven at 400 °C: (**a**) 12 A DC MS silver system exposed for 768 h, (**b**) HiPIMS 200 A silver-system exposed for 1800 h, and (**c**) 400 A HiPIMS silver system exposed for 384 h.

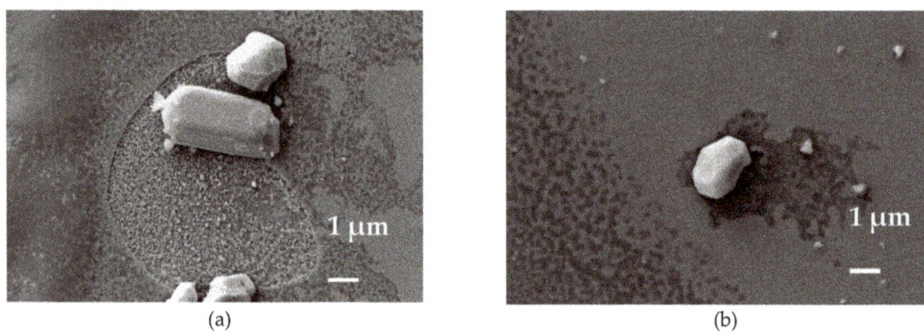

(a)   (b)

Figure 7. SEM of degradation on a protected silver system. (**a**) Taken from the 12 A DC MS silver system after 768 h at 400 °C. (**b**) Taken from the 200 A HiPIMS silver system after 768 h at 400 °C.

After 384 h, the 400 A HiPIMS silver system sample had visually degraded to be removed from the test. The nature of the degradation, although based on silver agglomeration, appeared different from the 12 A DC MS silver-based equivalent sample. The degradation patches for the 400 A HiPIMS based sample were smaller but occurred with higher frequency. In contrast, the 12 A DC MS samples exhibited larger corrosion patches, typified by the pronounced rupturing of the top barrier layer. The 12 A DC MS sample remained in the oven at 400 °C further until 768 h but, as can be observed from the photograph Figure 6a, showed significant degradation and was subsequently removed from the test. The HiPIMS 200 A, however, remained visually good aside from the edge effect and scratches from the polished substrate until the end of the testing period of 1800 h. The HiPIMS 200 A silver in this thin film system is thus shown to be the most durable in the 400 °C oven.

4. Discussion

The main aim of the work was to determine whether HiPIMS silver films, used in a protected thin film system, could improve the performance of high temperature durable reflectors in comparison to the standard DC MS equivalent. In this context, the effect of the HiPIMS mode and sputter-parameters was investigated on the resulting silver films in comparison to the 'baseline' DC MS silver used in protected silver systems.

As reported [7,13], the as-deposited HiPIMS silver exhibited lower deposition rates per power density compared to DC MS. This was partially accounted for, as the HiPIMS silver grown here were also denser as a function of increasing peak current. The increase in density with peak current is a well-reported phenomenon [14] and can be attributed to increased energy and thus mobility of ad-atoms due to an increased amount of ions in the deposition flux [15]. Also reported [16] but not measured here is that increasing the power density and ionized flux increases compressive stress in the silver films but without an increase in lattice defects.

The resistivity of the as-deposited HiPIMS silver films decreased slightly when increasing the peak target current from 100 A to 200A and was indeed lower (Figure 2b) than the 12 A DC MS silver. The decrease may be due to an increase in film density as a function of peak current [17].

The resistivity of the HiPIMS silver films, however, increased significantly when increasing further to 300 A and then 400 A to values higher than that of the reference 12 A DC MS silver. The resistivity is influenced by grain boundaries. A high number of grain boundaries and defects would reduce the efficiency of electron transport through the film, leading to a decrease in conductivity.

What was apparent was that the 200 A HiPIMS films had a similar grain size to the 12 A DC MS films. The grains had a higher degree of orientation and the film was much denser. The even denser 400 A HiPIMS film also showed a higher degree of grain orientation but a lower grain size. The film microstructure of the 400 A HiPIMS sample fits the extended Thornton model zone T [13] where the high energy leads to high surface diffusion and the formation of a dense, finer grained film with highly orientated grains. A high nucleation density, which leads to small grain size, has previously [18] been observed in HiPIMS and is enhanced by high ion currents towards the substrate surface. The increasing smaller grain size, exhibited at higher peak currents, thus increases the resistivity.

It seems that a competition between the two phenomena of the increasing density yet decreasing grain size occurred, where the highest conductivity under these sputter conditions occurred at 200 A in HiPIMS mode.

GIXRD and SEM showed that annealing the discrete silver films and the protected silver films led to grain growth and stronger orientation of the crystals. This, likewise, increased the conductivity for all the films presented here. Previous work [19] shows that above a certain time and temperature (and dependent on the initial thickness of the silver) the conductivity drastically reduces discrete silver films due to voiding and segregation of the silver into separated agglomerates. In this work here, the temperature and time for the relatively thick 500 nm silver did not exceed this point and the silver remained as a network; thus, only an increase in conductivity was measured upon annealing due to the increased grain size.

The critical parameter for the high temperature reflectors is not electrical conductivity but reflectance. The relationship between conductivity and reflectance for highly conducting metals, such as silver was derived by Drude and confirmed experimentally by Hagens-Ruben [20]. It was observed that, at higher wavelengths (lower frequencies), the optical constants of metals are similar to the values of Drude's function, where the complex refractive index is much smaller than the damping constant or extinction coefficient. This leads to high reflectance. At higher wavelengths (near infra-red to infra-red) one can thus partially correlate the increase (or decrease) in reflectance to the same physical reasons as the observed increase in conductivity (i.e., film densification) or decrease in conductivity (i.e., increase in grain boundaries due to smaller grain sizes). The as-deposited discrete silver films exhibited this correlation: the reflectivity increased with increasing film conductivity.

At higher frequencies, deviations of Drude's approach start to appear because bound electrons of the metal start to respond to the incidence of light instead of just valence band electron response. At 325 nm, there was a drop in the reflectivity of the films which correlated to an interband transition edge at 3.8 eV [21]. The broadening of this edge, observable in the measured reflectance spectra, with the highest current HiPIMS silver films may be correlated to defect energy states caused by, e.g., grain boundaries or other defect or impurities. Indeed, the biggest deviation in the solar reflectance between the sputtered silver samples was at this edge in the UV-visible portion of the spectrum. This is where the most gain was exhibited by the 200 A HiPIMS films. In the UV/visible portion of the spectrum absorption via surface plasmons [22], dependent on the surface geometry and size of the silver agglomerates on the surface, is also possible. This is particularly relevant for the discrete silver film post annealing. Clearly, from the SEM pictures of the annealed discrete silver films, the roughness increased and the surface geometry dramatically changed with annealing leading to the increased diffuse and decreased global solar reflectance.

For the protected silver film systems, the reflectance depended not only on the silver but also on the adhesion layers and barrier layers. As discussed previously in the results section, the benefits from the gain in the UV/visible portion of the spectra (around the interband transition edge) exhibited by the 200 A HiPIMS silver were not observable due to the interference effects and absorption of the metallic adhesion layer. The additional top barrier and adhesion layers were of course required as they limit the silver degradation. Thus, upon initially annealing, the protected silver film system grain growth occurred but without larger-scale agglomeration; thus, reflectance (and presumably the conductivity) increased.

Durability is defined in this context as minimized degradation in appearance and reflectance with increasing time at 400 °C. The 200 A HiPIMS silver system was more durable than the 12 A DC MS, whereas the 400 A HiPIMS silver system was less durable. The durability was determined by the whole film system, and it is beyond the scope of this paper to delve into the exact complex interplay between the film system and corrosion mechanism. The following aspects which affect the film stability are discussed: (1) The depositing flux during HiPIMS had a higher energy and higher ionized percentage, which would enhance the adhesion of the silver film to the substrate below, possibly improving durability. (2) The internal stress of the silver films, not measured here or even observed by a peak difference in the GIXRD, is reported [16] to be higher for the HiPIMS films, effecting the adhesion and interaction with the top barrier layer. (3) What was measured is a peak shift in the GIXRD spectra towards higher Bragg angles for all silver films in the protected film system after annealing. This shift, which was larger for the HiPIMS silver systems, implies an increase in compressive stress in the protected silver film with annealing. (4) Finally, the reduced grain size of the 400 A HiPIMS silver may be detrimental to the film stability (higher diffusion paths, increased free energy).

5. Conclusions

In conclusion, the 200 A HiPIMS yielded a denser layer than bulk silver films, resulting in a higher initial conductivity and reflectance than the reference DC MS sample. Annealing the silver film increased the conductivity (due to grain growth). In contrast, the reflectance decreased upon annealing due to surface structuring caused by silver agglomeration leading to unwanted light scattering and absorption by surface plasmons. The 200 A HiPIMS silver system was more durable at 400 °C than the DC MS silver equivalent system. Higher peak current HiPIMS yielded silver films which, despite the enhanced density, had smaller grains than the DC MS silver. This resulted in a decrease in conductivity and reflectance of the film. When the 400 A HiPIMS silver system was incorporated as part of a protected silver system it was less durable at 400 °C than the DC MS silver equivalent system. Thus HiPIMS, within a lower power density range which yields denser films yet does not exhibit a reduced grain size, is beneficial for the sputtered protected silver coating technology, in particular for the application in concentrated solar power secondary reflectors.

Supplementary Materials: The following are available online at http://www.mdpi.com/2079-6412/9/10/593/s1, Table S1: Data tabulated from the GIXRD spectra for the 400 A HiPIMS, 200 A HiPIMS, and 12 A DC magnetron sputtered silver shown in Figure 4a as discrete films on coated-glass substrates as-deposited and after an 8 h 400 °C anneal. Table S2: Data tabulated from the GIXRD spectra for the 400 HiPIMS, 200 HiPIMS, and 12 A DC magnetron sputtered silver shown in Figure 4b as protected silver film systems grown on coated glass substrates as deposited and after 24 h 400 °C anneal.

Author Contributions: Project conceptualization and supervision: C.H. and S.G., Layer investigation, experiment design, and analysis: S.G. and K.S., GIXRD measurement and analysis: C.W. and S.G., wrote the manuscript assisted by all other authors.

Funding: This research was funded by EU Horizon 2020 RAISELIFE, under grant agreement No. 686008.

Conflicts of Interest: The authors declare no conflict of interest.

References

1. Fernandez-Garcia, A.; Cantos-Soto, M.E.; Roger, M.; Wieckert, C.; Hutter, C.; Martinez-Arcos, L. Durability of solar reflector materials for secondary concentrators used in csp systems. *Sol. Energy Mater. Sol. Cells* **2014**, *130*, 51–63. [CrossRef]
2. Schöttl, P.; Zoschke, T.; Frantz, C.; Gilon, Y.; Heimsath, A.; Fluri, T. In Performance Assessment of a Secondary Concentrator for Solar Tower External Receivers. In Proceedings of the 24th Solar Paces Conference, Casablanca, Morocco, 2–5 October 2018.
3. Gudmundsson, J.T.; Brenning, N.; Lundin, D.; Helmersson, U. High power impulse magnetron sputtering discharge. *J. Vac. Sci. Technol. A* **2012**, *30*, 030801. [CrossRef]
4. West, G.T.; Kelly, P.J.; Bradley, J.W. A comparison of thin silver films grown onto zinc oxide via conventional magnetron sputtering and hipims deposition. *IEEE Trans. Plasma Sci.* **2010**, *38*, 3057–3061. [CrossRef]
5. Sarakinos, K.; Wordenweber, J.; Uslu, F.; Schulz, P.; Alami, J.; Wuttig, M. The effect of the microstructure and the surface topography on the electrical properties of thin ag films deposited by high power pulsed magnetron sputtering. *Surf. Coat. Technol.* **2008**, *202*, 2323–2327. [CrossRef]
6. Schneider, T.; Vucina, T.; Hee, C.A.; Araya, C.; Moreno, C. The gemini observatory protected silver coating: Ten years in operation. *Gr. Based Airborne Telesc.* **2016**, *9906*, 990632.
7. Anders, A. Deposition rates of high power impulse magnetron sputtering: Physics and economics. *J. Vac. Sci. Technol. A* **2010**, *28*, 783–790. [CrossRef]
8. Patterson, A. The scherrer formula for x-ray particle size determination. *Phys. Rev.* **1939**, *56*, 978–982. [CrossRef]
9. Sharma, S.K.; Spitz, J. Hillock formation, hole growth and agglomeration in thin silver films. *Thin Solid Films* **1980**, *65*, 339–350. [CrossRef]
10. Sharma, S.K.; Spitz, J. Hillock growth and agglomeration in thin silver films. *Thin Solid Films* **1979**, *61*, L13–L15. [CrossRef]
11. Grunwaldt, J.D.; Atamny, F.; Gobel, U.; Baiker, A. Preparation of thin silver films on mica studied by xrd and afm. *Appl. Surf. Sci.* **1996**, *99*, 353–359. [CrossRef]
12. Georg, A. Temperatur-korrosion an spiegelschichten. In *Jahrbuch Oberflächentechnik 2013*; Suchentrunk, R., Ed.; Leuze Verlag: Bad Saulgau, Germany, 2013; Volume 69, pp. 221–228.
13. Anders, A. A structure zone diagram including plasma-based deposition and ion etching. *Thin Solid Films* **2010**, *518*, 4087–4090. [CrossRef]
14. Alami, J.; Bolz, S.; Sarakinos, K. High power pulsed magnetron sputtering: Fundamentals and applications. *J. Alloy. Compd.* **2009**, *483*, 530–534. [CrossRef]
15. Petrov, I.; Adibi, F.; Greene, J.E.; Hultman, L.; Sundgren, J.E. Average energy deposited per atom—A universal parameter for describing ion-assisted film growth. *Appl. Phys. Lett.* **1993**, *63*, 36–38. [CrossRef]
16. Magnfalt, D.; Abadias, G.; Sarakinos, K. Atom insertion into grain boundaries and stress generation in physically vapor deposited films. *Appl. Phys. Lett.* **2013**, *103*, 051910. [CrossRef]
17. Choi, H.M.; Choi, S.K.; Anderson, O.; Bange, K. Influence of film density on residual stress and resistivity for cu thin films deposited by bias sputtering. *Thin Solid Films* **2000**, *358*, 202–205. [CrossRef]
18. Magnfalt, D.; Elofsson, V.; Abadias, G.; Helmersson, U.; Sarakinos, K. Time-domain and energetic bombardment effects on the nucleation and coalescence of thin metal films on amorphous substrates. *J. Phys. D Appl. Phys.* **2013**, *46*, 215303. [CrossRef]
19. Kim, H.C.; Alford, T.L.; Allee, D.R. Thickness dependence on the thermal stability of silver thin films. *Appl. Phys. Lett.* **2002**, *81*, 4287–4289. [CrossRef]

20. Silveira, F.E.M.; Kurcbart, S.M. Hagen-rubens relation beyond far-infrared region. *Eur. Phys. Lett* **2010**, *90*, 44004. [CrossRef]

21. Yang, H.H.U.; D'Archangel, J.; Sundheimer, M.L.; Tucker, E.; Boreman, G.D.; Raschke, M.B. Optical dielectric function of silver. *Phys. Rev. B* **2015**, *91*, 235137. [CrossRef]

22. Gong, J.B.; Dai, R.C.; Wang, Z.P.; Zhang, Z.M. Thickness dispersion of surface plasmon of ag nano-thin films: Determination by ellipsometry iterated with transmittance method. *Sci. Rep. UK* **2015**, *5*, 9279. [CrossRef] [PubMed]

MDPI
St. Alban-Anlage 66
4052 Basel
Switzerland
Tel. +41 61 683 77 34
Fax +41 61 302 89 18
www.mdpi.com

Coatings Editorial Office
E-mail: coatings@mdpi.com
www.mdpi.com/journal/coatings